Engineering BGM

CHAPMAN & HALL/CRC
Financial Mathematics Series

Aims and scope:
The field of financial mathematics forms an ever-expanding slice of the financial sector. This series aims to capture new developments and summarize what is known over the whole spectrum of this field. It will include a broad range of textbooks, reference works and handbooks that are meant to appeal to both academics and practitioners. The inclusion of numerical code and concrete real-world examples is highly encouraged.

Series Editors

M.A.H. Dempster
Centre for Financial Research
Judge Business School
University of Cambridge

Dilip B. Madan
Robert H. Smith School of Business
University of Maryland

Rama Cont
Center for Financial Engineering
Columbia University
New York

Published Titles

American-Style Derivatives; Valuation and Computation, *Jerome Detemple*

Financial Modelling with Jump Processes, *Rama Cont and Peter Tankov*

An Introduction to Credit Risk Modeling, *Christian Bluhm, Ludger Overbeck, and Christoph Wagner*

Portfolio Optimization and Performance Analysis, *Jean-Luc Prigent*

Robust Libor Modelling and Pricing of Derivative Products, *John Schoenmakers*

Structured Credit Portfolio Analysis, Baskets & CDOs, *Christian Bluhm and Ludger Overbeck*

Numerical Methods for Finance, *John A. D. Appleby, David C. Edelman, and John J. H. Miller*

Understanding Risk: The Theory and Practice of Financial Risk Management, *David Murphy*

Engineering BGM, *Alan Brace*

Proposals for the series should be submitted to one of the series editors above or directly to:
CRC Press, Taylor and Francis Group
24-25 Blades Court
Deodar Road
London SW15 2NU
UK

CHAPMAN & HALL/CRC FINANCIAL MATHEMATICS SERIES

Engineering BGM

Alan Brace

CRC Press is an imprint of the
Taylor & Francis Group, an **informa** business
A CHAPMAN & HALL BOOK

CRC Press
Taylor & Francis Group
6000 Broken Sound Parkway NW, Suite 300
Boca Raton, FL 33487-2742

First issued in paperback 2019

© 2008 by Taylor & Francis Group, LLC
CRC Press is an imprint of Taylor & Francis Group, an Informa business

No claim to original U.S. Government works

ISBN-13: 978-1-58488-968-7 (hbk)
ISBN-13: 978-0-367-38837-9 (pbk)

This book contains information obtained from authentic and highly regarded sources. Reasonable efforts have been made to publish reliable data and information, but the author and publisher cannot assume responsibility for the validity of all materials or the consequences of their use. The authors and publishers have attempted to trace the copyright holders of all material reproduced in this publication and apologize to copyright holders if permission to publish in this form has not been obtained. If any copyright material has not been acknowledged please write and let us know so we may rectify in any future reprint.

Except as permitted under U.S. Copyright Law, no part of this book may be reprinted, reproduced, transmitted, or utilized in any form by any electronic, mechanical, or other means, now known or hereafter invented, including photocopying, microfilming, and recording, or in any information storage or retrieval system, without written permission from the publishers.

For permission to photocopy or use material electronically from this work, please access www.copyright.com (http://www.copyright.com/) or contact the Copyright Clearance Center, Inc. (CCC), 222 Rosewood Drive, Danvers, MA 01923, 978-750-8400. CCC is a not-for-profit organization that provides licenses and registration for a variety of users. For organizations that have been granted a photocopy license by the CCC, a separate system of payment has been arranged.

Trademark Notice: Product or corporate names may be trademarks or registered trademarks, and are used only for identification and explanation without intent to infringe.

Visit the Taylor & Francis Web site at
http://www.taylorandfrancis.com

and the CRC Press Web site at
http://www.crcpress.com

Dedicated to the memory of my father George James Brace

Contents

Preface

Over the past several years the author has found himself frequently asked to give explanatory talks on BGM, some of which extended into one- or two-week workshops with detailed head-to-head technology transfer. The main interest came from small groups of quants either in banks or in software companies wanting to implement the model without wasting too much time decoding papers to find a suitable approach, and also academics and students wanting to get into the subject. This book is therefore naturally targeted at such people, who generally have several years experience around finance and a good grounding in the relevant mathematics.

The stimulus to begin writing was an invitation to join the Quantitative Finance Research Centre (QFRC) at the University of Technology Sydney (UTS) as an Adjunct Professor, and give a series of lectures on BGM for an audience of academics, students and industry quants over the course of a couple of semesters during 2006. This book grew out of those lectures, but the starting point was some eleven years of notes on various aspects of BGM, that were all prepared either for implementers writing production code, or as formal documentation to accompany production code, or in response to consulting tasks. Thus most of the techniques and methods described in this book originate in practical problems needing a solution and address real requirements. Moreover, many of them have been implemented, tried and tested either in an R&D environment like MatLab, or in production code.

A reader from a mathematics, physics or engineering background (or the quantitative end of another science) with a decent knowledge of analysis, optimization, probability and stochastic calculus (that is, familiar with *Ito* and *Girsanov* at the very least) should find this book fairly self-contained and thus hopefully a suitable resource and guide to implementing some version of the model. Indeed, part of the reason why the author has tried to keep the book relatively short is to make it easy to slip into one's briefcase and use as a ready reference; the other part is a pathological fear of catching *blitherer's disease*, which in extreme cases seems to dilute ideas to one per page!

The book starts with the standard lognormal flat BGM, and then focuses on the shifted (or displaced diffusion) version to develop basic ideas about construction, change of measure, correlation, calibration, simulation, timeslicing (like lattices), pricing, delta hedging, vega hedging, callable exotics and barriers. Further chapters cover cross-economy BGM, adaption of the HJM inflation model to the BGM framework, a simple tractable stochastic volatility version of BGM, and financial instruments in Brazil, which have evolved in a unique way and are amenable to BGM analysis.

Because shifted BGM can fit a cap or swaption implied volatility skew (but not a smile) and has the advantage of being just as tractable as flat BGM, it seems the right framework to present basic techniques. The stochastic volatility version aims to add a measure of convexity to the skew version, but we do not go so far as trying to calibrate to a full smile, which is a complex task appropriate to a cutting edge specialist. Overall the author can't help feeling that shifted BGM with the stochastic volatility extension as described here is about right for both the Mortgage Backed Security world, and also second tier banks wanting a robust framework in which to manage structured products sold into their customer base, without having to worry too much about being arbitraged.

To sum up, the reader is presented with several, progressively more sophisticated, versions of BGM, and a range of methods and recipes that (after some expansion and articulation) can be programmed into production code, and is free to choose an implementation to suit his requirements. Thus the book attempts to be an *implementer's handbook* offering straightforward models suitable for more conservative institutions who want a robust, safe and stable environment for calibrating, simulating, pricing and hedging interest rate instruments. Advanced versions for market makers, hedge funds or leading international banks are left to their top quants, though their newer quants might conveniently learn about market models from this book and then do better.

Many people contributed in some way to this book. In particular, it was a pleasure working with Marek Musiela through the early '90s at Citibank, where Mike Hawker in Sydney and Pratap Sondhi in Hong Kong provided support and a framework to do much of the original work. Since then, innumerable conversations with colleagues, reading and decoding many excellent papers, attendance at wide ranging professional conferences and some foolish mistakes have added enormously to the author's basic knowledge.

In direct preparation of this manuscript Chapman and Hall were patient and encouraging, Marek Rutkowski gave me a copy of his extensive bibliography greatly simplifying the task of preparing references, and my thanks to Carl Ang, Peter Buchen, Andrew Campbell, Daniel Campos, Tim Glass, Ben Goldys, Ivan Guo, Steve McCarthy, Frank Merino, Paul O'Brien, Erik Schlogl and Rob Womersley for helping check different parts of the book. Further thanks are due to both National Australia Bank[1] and UTS for their material support in terms of time and infrastructure over the past couple of years, and also to *MY* for encouragement at some difficult moments.

A word on the title 'Engineering BGM'. The background is that Miltersen, Sandmann and Sondermann (MSS), see [78], were the first to get a 'kosher'

[1]All views expressed in this book are the author's and in no way reflect NAB policy, philosophy or technology.

Black caplet formula out of HJM, but unfortunately they did not establish existence, which is an essential feature of a model (along with, the author feels, the technology to price complex options). We, that is Brace, Gatarek and Musiela (BGM), see [30], grasped the intuition behind the model, proved existence, derived swaption formulae, calibrated to the market and constructed simulation technology for pricing.

So generally speaking the model has more-or-less become known as 'BGM' in the industry and the 'Libor Market Model' in academic circles. My preference for the title 'Engineering BGM' over the alternative 'Engineering the Libor Market Model', is partly because this book is aimed at industry quants and traders and partly because it is shorter and more punchy. But unequivocally, MSS made the first breakthrough in this area, and we referenced their work in our paper [30] describing it as a 'key piece of information'.

Finally, if that nightmare for a single author 'the bad stupid mistake' should materialize, it is soley the author's fault and he apologizes in advance. Of course, all information about any, hopefully more minor, mistakes found by readers would be gratefully received (at any one of the author's email addressess on the title page), as would any suggestions for inclusions, exclusions and better ways of doing things (in case there should ever be a second edition of this book).

Alan Brace

(Sydney 25 September 2007)

Chapter 1

Introduction

Modern interest rate modelling began[1] with Ho & Lee's (HL) important 1986 paper [54], and matured into the Heath, Jarrow and Morton (HJM) model [52], which was circulating in 1988, and which became the standard framework for interest rates in the early '90s. Initial work on the market models was done within that framework, so to set the scene, the single-currency *domestic* version of HJM is reviewed in Section-1.1.

When the volatility function is deterministic, HJM is Gaussian, extremely tractable, and includes versions like Hull and White [58] and many other models. But until the advent of the *market models* [30], [66], [78] and [79] around 1994-97, the market's use of the *Black caplet* and *Black swaption* formulae (which priced assuming that forwards and swaprates were lognormal) was regarded as an aberration which could not be reconciled with HJM. A further problem was that HJM *exploded* when the instantaneous forward rates were made lognormal. The author can recall comments at conferences in the early '90s along the lines that 'the market is foolish and should adopt some arbitrage free Gaussian HJM model as a standard'.

To avoid explosions, attention shifted to modelling the cash forwards, and in 1994 Miltersen, Sandmann and Sondermann [78] found a PDE method, described in Section-1.2 below, to derive the Black caplet formula within the arbitrage free HJM framework. Knowing that was possible, and that the Black caplet formula was not an aberration, was a key piece of information.

The author's main contribution to events was to grasp the intuition, described in Section-1.3 below, that the cash forwards *want to be lognormal*, but under the forward measure at the *end of their interval.* With that realization, the derivation of the Black caplet formula became trivial, and led to the so called *forward construction* of BGM detailed in [30], by Musiela, Gatarek and the author, which established existence of the model, derived approximate analytic swaption formulae, calibrated to the market, and provided suitable simulation technology for pricing exotics.

[1] Though intriguingly, the previous long standing actuarial practice of hedging bonds by matching duration turned out to be equivalent to delta hedging within the HL model.

1.1 Background HJM

REMARK 1.1 Before beginning, a word on our '$*$' notation for transposes. Throughout this book we will generally be dealing with multi-factor models involving an n-dimensional vector volatility function, say $\xi : \mathbb{R} \to \mathbb{R}^n$ and a corresponding multi-dimensional Brownian motion $W(t) \in \mathbb{R}^n$. Usually they (or similar expressions as in (1.3) below) appear together as inner products, so we use the '$*$' notation to indicate transpose and write

$$\xi^*(t)\, dW(t) \equiv \langle \xi(t), dW(t) \rangle$$

for that inner product. Of course, in single factor models $\xi(t)\, dW(t)$ would simply mean the product of two scalor quantities. Note that many authors today adopt the practice (which is beginning to appeal to the author) of simply writing $\xi(t)\, dW(t)$ and leaving the reader to work out from the context if an inner product is implied. □

The *ingredients* of the HJM domestic interest rate model are:

1. An *instantaneous* at t *forward rate* $f(t,T)$ for maturity T, with SDE

$$df(t,T) = \alpha(t,T)\, dt + \sigma^*(t,T)\, dW_0(t) \tag{1.1}$$

   where the stochastic driving variable $W_0(t)$ is multi-dimensional Brownian motion *(BM)* under the *arbitrage-free measure* $\mathbb{P}_0$, and $\sigma(t,T)$ is a possibly stochastic vector *volatility function* for $f(t,T)$.

2. A *spot rate* $r(t) = f(t,t)$ and *numeraire bank account* to accumulate it

$$\beta(t) = \exp\left(\int_0^t r(s)\, ds\right),$$

3. *Assets* in the form of a spectrum of time T maturing *zero coupon* bonds

$$B(t,T) = \exp\left(-\int_t^T f(t,u)\, du\right),$$

   paying 1 at their maturity T.

To be arbitrage free, the zeros discounted by the bank account as numeraire

$$Z(t,T) = \frac{B(t,T)}{\beta(t)} = \exp\left(-\int_0^t r(s)\, ds - \int_t^T f(t,u)\, du\right), \tag{1.2}$$

must be $\mathbb{P}_0$-martingales for all T. Because

$$d\int_t^T f(t,u)\,du = \int_t^T df(t,u)\,du - f(t,t)\,dt,$$
$$= -r(t)\,dt + \left(\int_t^T \alpha(t,u)\,du\right)dt + \left(\int_t^T \sigma^*(t,u)\,du\right)dW_0(t),$$

applying Ito to (1.2), the SDE for $Z(t,T)$ is

$$\frac{dZ(t,T)}{Z(t,T)} = \left\{ \begin{array}{c} -r(t)\,dt + r(t)\,dt - \left(\int_t^T \alpha(t,u)\,du\right)dt \\ -\left(\int_t^T \sigma^*(t,u)\,du\right)dW_0(t) + \frac{1}{2}\left|\int_t^T \sigma(t,u)\,du\right|^2 dt \end{array} \right\},$$
$$= \left\{ \begin{array}{c} -\left[\int_t^T \alpha(t,u)\,du - \frac{1}{2}\left|\int_t^T \sigma(t,u)\,du\right|^2\right]dt \\ -\int_t^T \sigma^*(t,u)\,du\,dW_0(t) \end{array} \right\}.$$

For this to be a $\mathbb{P}_0$ martingale the drift must vanish, so

$$\alpha(t,T) = \sigma^*(t,T)\int_t^T \sigma(t,u)\,du,$$

and the SDE for the instantaneous forwards is

$$df(t,T) = \sigma^*(t,T)\int_t^T \sigma(t,u)\,du\,dt + \sigma^*(t,T)\,dW_0(t). \tag{1.3}$$

Differentiating $B(t,T) = \beta(t)\,Z(t,T)$, the corresponding SDE for the zero coupon bond is

$$\frac{dB(t,T)}{B(t,T)} = r(t)\,dt - \int_t^T \sigma^*(t,u)\,dudW_0(t). \tag{1.4}$$

REMARK 1.2 The HJM approach therefore implies that the volatility

$$b(t,T) = -\int_t^T \sigma(t,u)\,du, \tag{1.5}$$

of each zero coupon bond $B(t,T)$ is continuous in T, a restriction ruling out piecewise constant bond volatilities. □

Because assets discounted by the bank account numeraire are $\mathbb{P}_0$-martingales, the *present value* of a cashflow $X(T)$ occurring at time T is

$$X(t) = \mathbf{E}_0\left(\left.\frac{\beta(t)}{\beta(T)}X(T)\right|\mathcal{F}_t\right), \tag{1.6}$$

where $\mathbf{E}_0$ is expectation under $\mathbb{P}_0$, and $\mathcal{F}_t$ is the underlying filtration (total accumulated information up to t). In particular, because a zero coupon pays 1 at maturity

$$B(t,T) = \mathbf{E}_0\left(\frac{\beta(t)}{\beta(T)}1\middle|\mathcal{F}_t\right) = \mathbf{E}_0\left(\exp\left(-\int_t^T r(s)\,ds\right)\middle|\mathcal{F}_t\right). \tag{1.7}$$

A *forward contract* $F_T(t,T_1)$ on a zero-coupon bond $B(t,T_1)$ maturing at T_1, exchanges at time T the zero coupon $B(T,T_1)$ for $F_T(t,T_1)$. The present value of the exchange must be zero, hence $F_T(t,T_1)$ must satisfy

$$\mathbf{E}_0\left\{\frac{\beta(t)}{\beta(T)}\left[F_T(t,T_1) - B(T,T_1)\right]\middle|\mathcal{F}_t\right\} = 0$$

giving the following *model free result* for forward contracts

$$F_T(t,T_1) = \frac{B(t,T_1)}{B(t,T)}. \tag{1.8}$$

When $T_1 = T+\delta$, the *cash forward* $K(t,T)$ over the interval $(T,T_1]$ is defined in terms of the forward contract $F_T(t,T_1)$ by

$$F_T(t,T_1) = \frac{B(t,T_1)}{B(t,T)} = \frac{1}{1+\delta K(t,T)}. \tag{1.9}$$

REMARK 1.3 In the following equation (1.10), please note that the one variable Radon-Nikodym derivative $Z(t) = \mathbf{E}_0\{Z(T)|\,\mathcal{F}_t\}$ is not the two variable discounted zero coupon function $Z(t,T) = \frac{B(t,T)}{\beta(t)}$. □

Being a strictly positive process, the bank account $\beta(t)$ induces a *forward measure* $\mathbb{P}_T$ (expectation $\mathbf{E}_T$) at any maturity T through

$$\mathbb{P}_T = Z_T\,\mathbb{P}_0 \quad or \quad \mathbf{E}_T\{\cdot\} = \mathbf{E}\{\cdot\; Z_T\} \tag{1.10}$$
$$Z(T) = \frac{1}{\beta(T)\,B(0,T)}.$$

It follows, from the conditional change of measure result of Appendix-A.3.5, that

$$\mathbf{E}_T(X(T)|\,\mathcal{F}_t) = \frac{\mathbf{E}_0(X(T)\,Z(T)|\,\mathcal{F}_t)}{\mathbf{E}_0(Z(T)|\,\mathcal{F}_t)} = \frac{\mathbf{E}_0\left(\frac{\beta(t)}{\beta(T)}X(T)\middle|\,\mathcal{F}_t\right)}{\mathbf{E}_0\left(\frac{\beta(t)}{\beta(T)}\middle|\,\mathcal{F}_t\right)} = \frac{X(t)}{B(t,T)},$$

which simplifies the present value equation (1.6) to

$$X(t) = \mathbf{E}_0\left(\frac{\beta(t)}{\beta(T)}X(T)\middle|\,\mathcal{F}_t\right) = B(t,T)\,\mathbf{E}_T(X(T)|\,\mathcal{F}_t). \tag{1.11}$$

Also $X(t)$ discounted by $B(t,T)$ is a martingale under the forward measure $\mathbb{P}_T$ because for $s<t$

$$\mathbf{E}_T\left(\left.\frac{X(t)}{B(t,T)}\right|\mathcal{F}_s\right)=\mathbf{E}_T\left(\mathbf{E}_T\left(X(T)|\,\mathcal{F}_t\right)|\,\mathcal{F}_s\right)=\mathbf{E}_T\left(X(T)|\,\mathcal{F}_s\right)=\frac{X(s)}{B(s,T)}.$$

Integrating (1.4) over $[0,T]$ identifies $Z(T)$ because

$$B(T,T)=1=B(0,T)\,\beta(T)\,\mathcal{E}\left(-\int_0^T\int_t^T\sigma^*(t,u)\,du\,dW_0(t)\right),$$

$$\Rightarrow\qquad Z(T)=\mathcal{E}\left\{-\int_0^T\int_t^T\sigma^*(t,u)\,du\,dW_0(t)\right\},$$

showing, from the Girsanov Theorem of Section-A.3.5, that $W_T(t)$, given by

$$dW_T(t)=dW_0(t)+\int_t^T\sigma(t,u)\,du\;dt, \tag{1.12}$$

is $\mathbb{P}_T$-BM. Subtracting from a similar expression for $W_{T_1}(t)$, a $\mathbb{P}_{T_1}$-BM,

$$dW_{T_1}(t)=dW_T(t)+\int_T^{T_1}\sigma(t,u)\,du\;dt. \tag{1.13}$$

From equations (1.4), (1.9) and the result in the Appendix A.3.3, the SDE for the forward contract $F_T(t,T_1)$ is

$$\begin{aligned}\frac{dF_T(t,T_1)}{F_T(t,T_1)}&=\left\{\begin{array}{c}r(t)\,dt-r(t)\,dt\\-\int_T^{T_1}\sigma^*(t,u)\,du\left[dW_0(t)+\int_t^T\sigma(t,u)\,du\;dt\right]\end{array}\right\},\\&=-\int_T^{T_1}\sigma^*(t,u)\,du\,dW_T(t),\end{aligned} \tag{1.14}$$

while the SDE for its reciprocal is

$$\begin{aligned}\frac{d\left(\frac{1}{F_T(t,T_1)}\right)}{\left(\frac{1}{F_T(t,T_1)}\right)}&=\left\{\begin{array}{c}r(t)\,dt-r(t)\,dt\\+\int_T^{T_1}\sigma^*(t,u)\,du\left[dW_0(t)+\int_t^{T_1}\sigma(t,u)\,du\;dt\right]\end{array}\right\},\\&=\int_T^{T_1}\sigma^*(t,u)\,du\,dW_{T_1}(t).\end{aligned} \tag{1.15}$$

Hence $F_T(t,T_1)$ is a $\mathbb{P}_T$-martingale while, more importantly as we will see, its reciprocal $\frac{1}{F_T(t,T_1)}$ is a $\mathbb{P}_{T_1}$-martingale.

1.2 The first 'correct' Black caplet

Miltersen, Sandmann and Sondermann [78] started with the assumption that under the T-forward measure $\mathbb{P}_T$ the cash forward $K(t,T)$ over $[T,T_1]$ was of *lognormal type* with deterministic volatility γ (which we here set constant for easy exposition), that is, they assumed the SDE for $K(t,T)$ has form

$$dK(t,T) = (drift)\,dt + K(t,T)\;\gamma\; dW_T(t)\,, \tag{1.16}$$

and then worked with the corresponding forward contract $F_T(t,T_1)$ (because it is a $\mathbb{P}_T$-martingale). Differentiating (1.9) using (1.14), and then comparing the stochastic term with that of (1.16), gives an SDE for $F_T(t,T_1)$:

$$\begin{aligned} dK(t,T) &= \frac{1}{\delta} d\left(\frac{1}{F_T(t,T_1)} - 1\right) \\ &= (drift)\,dt + \frac{1}{\delta F_T(t,T_1)} \int_T^{T_1} \sigma(t,u)\,du\; dW_T(t) \\ \Rightarrow \quad & \int_T^{T_1} \sigma(t,u)\,du = K(t,T)\,\gamma\delta F_T(t,T_1) = [1 - F_T(t,T_1)]\,\gamma \\ \Rightarrow \quad & dF_T(t,T_1) = -F_T(t,T_1)\,[1 - F_T(t,T_1)]\;\gamma\; dW_T(t)\,. \end{aligned} \tag{1.17}$$

The time t value of a Black caplet struck at κ, fixed at T and paid at T_1, is

$$\begin{aligned} \mathrm{cpl}(t) &= \mathbf{E}_0 \left\{ \frac{1}{\beta(T_1)} \delta \left[K(T,T) - \kappa\right]^+ \middle| \mathcal{F}_t \right\} \\ &= \mathbf{E}_0 \left\{ \frac{B(T,T_1)}{\beta(T)} \delta \left[K(T,T) - \kappa\right]^+ \middle| \mathcal{F}_t \right\}, \\ &= B(t,T)\,\mathbf{E}_T \left\{ F_T(T,T_1) \left[\frac{1}{F_T(T,T_1)} - 1 - \delta\kappa\right]^+ \middle| \mathcal{F}_t \right\}, \\ &= B(t,T)\,\mathbf{E}_T \left\{ \left[1 - (1+\delta\kappa)\,F_T(T,T_1)\right]^+ \middle| \mathcal{F}_t \right\}. \end{aligned}$$

Applying Ito, Miltersen et al then set to zero the drift of the $\mathbb{P}_T$-martingale

$$v\left(t, F_T(t,T_1)\right) = \frac{\mathrm{cpl}(t)}{B(t,T)},$$

so that $\mathrm{cpl}(t)$ is given by the solution $v\left(t, F_T(0,T_1)\right)$ to the non-linear PDE

$$\frac{\partial v}{\partial t} + \tfrac{1}{2}\gamma^2 x^2 (1-x)^2 \frac{\partial^2 v}{\partial x^2} = 0 \qquad \textit{with} \qquad v(T,x) = \left[1 - (1+\delta\kappa)\,x\right]^+.$$

This converts to a *heat equation* problem with the transformations

$$s = \gamma^2 (T-t), \quad z = \ln \frac{x}{1-x}, \tag{1.18}$$

$$v(t,x) = \frac{e^{-\frac{s}{8}}}{e^{\frac{z}{2}} + e^{-\frac{z}{2}}} u(s,z) \Rightarrow \quad \frac{\partial u}{\partial s} = \frac{1}{2}\frac{\partial^2 u}{\partial z^2}$$

$$\textit{with} \quad u(0,z) = \left\{e^{\frac{z}{2}} + e^{-\frac{z}{2}}\right\}\left[1 - \frac{(1+\delta\kappa)}{1+e^{-z}}\right]^+,$$

which has the solution (substitute in the PDE and integrate by parts)

$$\begin{aligned} u(s,z) &= \int_{-\infty}^{\infty} u\left(0, z + v\sqrt{s}\right) \mathbf{N}_1(v)\, dv \\ &= \int_{-\infty}^{\Upsilon} \left[\exp\left(-\tfrac{1}{2}\left[z + v\sqrt{s}\right]\right) - \delta\kappa \exp\left(\tfrac{1}{2}\left[z + v\sqrt{s}\right]\right)\right] \mathbf{N}_1(v)\, dv, \\ &= \exp\left(-\frac{z}{2} + \frac{s}{8}\right)\mathbf{N}\left(\Upsilon + \frac{\sqrt{s}}{2}\right) - \delta\kappa \exp\left(\frac{z}{2} + \frac{s}{8}\right)\mathbf{N}\left(\Upsilon - \frac{\sqrt{s}}{2}\right), \\ &\textit{in which} \quad \Upsilon = -\frac{1}{\sqrt{s}}(z + \ln \delta\kappa). \end{aligned}$$

Inverting the transforms (1.18) to go from $u(s,z)$ back to $v(t,x)$

$$v(t,x) = x\left\{\frac{[1-x]}{x}\mathbf{N}(h) - \delta\kappa\mathbf{N}\left(h - \tfrac{1}{2}\gamma\sqrt{T-t}\right)\right\},$$

$$h = \frac{1}{\gamma\sqrt{T-t}}\left\{\ln\left(\frac{1-x}{x}\frac{1}{\delta\kappa}\right) + \tfrac{1}{2}\gamma^2(T-t)\right\}$$

the caplet price $\mathrm{cpl}(t)$ follows from $v(t,x)$ on using

$$x = F_T(t,T_1) = \frac{B(t,T_1)}{B(t,T)}, \qquad \delta K(t,T) = \frac{1-x}{x},$$

$$\mathrm{cpl}(t) = B(t,T)\, v(t,x) = \delta B(t,T_1)\ \mathbf{B}\left(K(t,T),\ \kappa,\ \gamma\sqrt{T-t}\right).$$

where $\mathbf{B}(\cdot)$ is the Black formula, see Appendix-A.2.3.

A probabilistic proof of this result obtained by the author while trying to articulate the insight of MSS, runs as follows. Simplify notation by setting $\mathbb{P}_T = \mathbb{P}$, $F_T(t,T_1) = F_t$, $K_t = K(t,T)$ and $W_T(t) = W_t$. From the SDE (1.17) for F_t, if

$$Z_t = \ln\frac{F_t}{1-F_t} \quad \textit{or} \quad F_t = \frac{1}{1+\exp(-Z_t)}, \quad \textit{then}$$

$$dZ_t = -\gamma\left[dW_t - \tfrac{1}{2}\gamma \tanh\left(\tfrac{1}{2}Z_t\right)dt\right], \quad \exp(-Z_t) = \delta K_t.$$

Change measure between $\mathbb{P}$ with BM W_t, and $\mathbb{Q}$ with BM $\widetilde{W}_t$ according to

$$dW_t = d\widetilde{W}_t + \tfrac{1}{2}\gamma \tanh\left(\tfrac{1}{2}Z_t\right) dt$$
$$\Rightarrow \quad dZ_t = -\gamma \widetilde{W}_t, \quad \langle Z \rangle_t = \gamma^2 t \quad and \quad Z_T - Z_t = -\gamma\left(\widetilde{W}_T - \widetilde{W}_t\right),$$

which, from Girsanov's theorem (A.3.5), means

$$\begin{aligned}\mathbb{P} &= \mathcal{E}\left\{-\int_0^T \tfrac{1}{2}\gamma \tanh\left(\tfrac{1}{2}Z_t\right) d\widetilde{W}_t\right\} \mathbb{Q} = \mathcal{E}\left\{\int_0^T \tfrac{1}{2}\tanh\left(\tfrac{1}{2}Z_t\right) dZ_t\right\} \mathbb{Q} \\ &= \exp\left(-\tfrac{1}{8}\gamma^2 [T-t]\right) \frac{\cosh\left(\frac{1}{2}Z_T\right)}{\cosh\left(\frac{1}{2}Z_t\right)} \mathbb{Q}\end{aligned}$$

because

$$d\left[\ln\cosh\left(\tfrac{1}{2}Z_t\right)\right] = \tfrac{1}{2}\tanh\left(\tfrac{1}{2}Z_t\right) dZ_t + \tfrac{1}{8}\operatorname{sech}^2\left(\tfrac{1}{2}Z_t\right) d\langle Z\rangle_t \quad \Rightarrow$$
$$\mathcal{E}\left\{\int_t^T \tfrac{1}{2}\tanh\left(\tfrac{1}{2}Z_s\right) dZ_s\right\} = \exp\left\{\int_t^T d\left[\ln\cosh\left(\tfrac{1}{2}Z_s\right)\right] - \tfrac{1}{8}\int_t^T d\langle Z\rangle_s\right\}.$$

Hence, using (A.3.4) and (A.2.3), the time-t value of the option is

$$\begin{aligned}\operatorname{cpl}(t) &= B(t,T)\,\mathbf{E}_{\mathbb{P}}\left\{\left[1-(1+\delta\kappa)F_T\right]^+ \middle| \mathcal{F}_t\right\}, \\ &= B(t,T)\,\mathbf{E}_{\mathbb{Q}}\left\{\exp\left(-\tfrac{1}{8}\gamma^2[T-t]\right)\tfrac{\cosh\left(\frac{1}{2}Z_T\right)}{\cosh\left(\frac{1}{2}Z_t\right)}\left[1-\tfrac{(1+\delta\kappa)}{1+\exp(-Z_T)}\right]^+ \middle| \mathcal{F}_t\right\}, \\ &= \tfrac{B(t,T)}{[1+\exp(-Z_t)]}\mathbf{E}_{\mathbb{Q}}\left\{\begin{bmatrix}\exp(-Z_t)\exp\left(-\frac{1}{2}[Z_T-Z_t]-\frac{1}{8}\gamma^2[T-t]\right) \\ -\delta\kappa\exp\left(\frac{1}{2}[Z_T-Z_t]-\frac{1}{8}\gamma^2[T-t]\right)\end{bmatrix}^+ \middle| \mathcal{F}_t\right\}, \\ &= \delta B(t,T)\,F_T(t,T_1)\,\mathbf{E}_{\mathbb{Q}}\left\{\begin{bmatrix}K_t\mathcal{E}\left(\frac{1}{2}\gamma\left(\widetilde{W}_T-\widetilde{W}_t\right)\right) \\ -\kappa\mathcal{E}\left(-\frac{1}{2}\gamma\left(\widetilde{W}_T-\widetilde{W}_t\right)\right)\end{bmatrix}^+ \middle| \mathcal{F}_t\right\}, \\ &= \delta B(t,T_1)\,\mathbf{B}\left(K(t,T),\ \kappa,\ \gamma\sqrt{T}\right).\end{aligned}$$

1.3 Forward BGM construction

The *intuition behind BGM* is that the forward $K(t,T)$ over the interval $(T,T_1]$ wants to be lognormal, but under the forward measure $\mathbb{P}_{T_1}$ located at its payoff T_1 at the *end of the interval.* Specifically, recall (1.9) that the cash forward $K(t,T)$ over $(T,T_1]$ with *coverage* $\delta = |(T,T_1]|$ is related to the reciprocal of the forward contract by

$$1 + \delta K(t,T) = \frac{1}{F_T(t,T_1)}.$$

Differentiating this equation using (1.15) gives

$$dK(t,T)\ \delta = [1+\delta K(t,T)] \int_T^{T_1} \sigma^*(t,u)\,du\,dW_{T_1}(t).$$

So if the HJM volatility function $\sigma(t,T)$ is made stochastic and chosen to satisfy

$$\int_T^{T_1} \sigma(t,u)\,du = \frac{\delta K(t,T)}{1+\delta K(t,T)}\xi(t,T), \tag{1.19}$$

where $\xi(t,T)$ is a deterministic vector function, then the forward $K(t,T)$ becomes lognormal under the $\mathbb{P}_{T_1}$-forward measure

$$dK(t,T) = K(t,T)\,\xi^*(t,T)\ dW_{T_1}(t).$$

That produces the *Black formula* (A.2.3) for the time t value $\operatorname{cpl}(t)$ of a caplet struck at κ, setting at T and paying at T_1, because

$$\begin{aligned}
\operatorname{cpl}(t) = \operatorname{cpl}(t,\kappa,T,T_1) &= \mathbf{E}_0\left\{ \frac{\beta(t)}{\beta(T_1)}\delta\left[K(T,T)-\kappa\right]^+ \middle| \mathcal{F}_t \right\}, \qquad (1.20)\\
&= \delta B(t,T_1)\,\mathbf{E}_{T_1}\left\{ \left[K(t,T)\,\mathcal{E}\left(\int_t^T \xi^*(s,T)\ dW_{T_1}(s) \right) - \kappa \right]^+ \middle| \mathcal{F}_t \right\},\\
&= \delta B(t,T_1)\,\mathbf{B}\left(K(t,T),\ \kappa,\ \sqrt{\int_t^T |\xi(s,T)|^2\,ds} \right).
\end{aligned}$$

In general, for $t > T_j$ the forwards $K(t,T_j)$ are *dead* and equal to $K(T_j,T_j)$, while for $t \le T_j$ the forwards $K(t,T_j)$ are *alive* and lognormal under $\mathbb{P}_{T_{j+1}}$. By repeatedly using (1.13) and (1.19) for $t \le T_j$ SDEs for any $K(t,T_j)$ under one fixed *terminal measure* $\mathbb{P}_n$ at T_n $(n > j)$ are therefore

$$\frac{dK(t,T_j)}{K(t,T_j)} = \begin{array}{l} -\sum_{k=j+1}^{n-1} \dfrac{\delta K(t,T_k)}{1+\delta K(t,T_k)}\xi^*(t,T_k)\,\xi(t,T_j)\,dt \\ +\xi^*(t,T_j)\,dW_n(t). \end{array} \tag{1.21}$$

That provides a framework for simulation because $\mathbb{P}_n$ is a unique measure under which the random number driving the simulation can be regarded as working.

Establishing existence of this *forward constructed* BGM model, however, is not straightforward because the HJM volatility function $\sigma(t,T)$ cannot be recovered from (1.19) in terms of the variables t,T and $f(t,T)$, making it impossible to return to (1.3) to query existence.

Thus a direct proof of the existence of a solution to the system of equations (1.21) is required, and in the original BGM paper [30] it was pushed through by letting all coverages be equal at δ (thereby creating a continuous spectrum of forwards $K(t,T)$ for all maturities), setting $\sigma(t,T) = 0$ for $t \leq T \leq t + \delta$, and then working in the setting of infinite dimensional SDEs.

REMARK 1.4 An alternative way to construct BGM is suggested by the technique for jointly simulating the SDEs (1.21) under $\mathbb{P}_n$. The expression for the drift in the SDE for $K(t,T_j)$ contains only later maturity forwards $K(t,T_{j+1})$,...,$K(t,T_{n-1})$, so one must work backwards $j = n-2, n-3, ...$ obtaining successive incremented forwards $K(t+\Delta t, T_j)$ from the later maturing forwards. The intuition that *capacity to program* is close to *proof by construction,* then suggests the *backward construction* approach of Section-3.2. □

Chapter 2

Bond and Swap Basics

The most basic of interest rate instruments are bonds and swaps, and in one form or another they underlie most activity in interest rate trading.

The notion of a zero coupon bond, an asset paying 1 at its maturity, was introduced in Section-1.1. The more usual *coupon bond*, which pays a series of usually semiannual coupons before returning the principal 1 at maturity, can of course be expressed as a linear combination of zero-coupon bonds, but they are not really within the scope of this book, and we do not consider them.

The basic object modelled in HJM - the instantaneous forward rate - is not observed in the market, so working with it requires abstraction through at least one intermediate layer to reach real world objects like bonds or options. Thus, for example, calibrating HJM to options, that is, fitting the instantaneous forward rate volatility function $\sigma(t,T)$ to option implied volatilities, can be unnecessarily complicated. Moreover, see Remark-1.2, HJM bond volatilities must be continuous, which is restrictive.

For these reasons, in the first section, we develop some arbitrage free methods based on zero-coupon bonds and forward measures to tackle construction of shifted BGM by *backward induction.*

The other fundamental interest rate instrument is the *swap*, in which interest rate risk is exchanged between two parties by swapping a sequence of *floating Libor* rates against a series of *fixed coupons* (which contrarily may actually be complex and variable in some exotic swaps).

Thus, in the second section, we derive some fundamental model free results for swaps, while establishing a standard swap notation, summarized in Section-A.1, that we will adopt (as far as possible) throughout this book.

2.1 Zero coupon bonds - drifts and volatilities

We will assume that *zero coupon* bonds $B(t,T)$ follow Ito processes like

$$\frac{dB(t,T)}{B(t,T)} = f(t)\,dt + b^{*}(t,T)\,dW(t) \quad t \le T, \tag{2.1}$$

under some fixed reference measure $\mathbb{P}$ (like the *real world* measure). Note that while the bond volatility vector $b(t,T)$ is maturity dependent and in general different for each bond, the drift $f(t)$ is assumed maturity independent and the same for all bonds.

To justify that suppose the contrary were true and

$$\frac{dB(t,T)}{B(t,T)} = f(t,T)\,dt + b^*(t,T)\,dW(t).$$

Presuming just a 2-factor model (the following argument works equally well for n-factors), arbitrarily select 3 bonds maturing at different T_1, T_2 and T_3, and eliminate the risky, that is, stochastic, components between them. That produces the following expression involving two determinants

$$\begin{vmatrix} \frac{dB(t,T_1)}{B(t,T_1)} & \frac{dB(t,T_2)}{B(t,T_2)} & \frac{dB(t,T_3)}{B(t,T_3)} \\ b^{(1)}(t,T_1) & b^{(1)}(t,T_2) & b^{(1)}(t,T_3) \\ b^{(2)}(t,T_1) & b^{(2)}(t,T_2) & b^{(2)}(t,T_3) \end{vmatrix} = \begin{vmatrix} f(t,T_1) & f(t,T_2) & f(t,T_3) \\ b^{(1)}(t,T_1) & b^{(1)}(t,T_2) & b^{(1)}(t,T_3) \\ b^{(2)}(t,T_1) & b^{(2)}(t,T_2) & b^{(2)}(t,T_3) \end{vmatrix} dt. \tag{2.2}$$

The right-hand side of equation (2.2) is risk free, implying that if we took *fixed at time t constant amounts*

$$A_i = \frac{1}{B(t,T_i)} \begin{vmatrix} b^{(1)}(t,T_j) & b^{(1)}(t,T_k) \\ b^{(2)}(t,T_j) & b^{(2)}(t,T_k) \end{vmatrix} \quad \textit{of bond} \quad B(t,T_i),$$
$$\textit{for} \qquad (i,j,k) = (1,2,3), \quad (2,3,1) \quad \textit{and} \quad (3,1,2)$$

the resulting portfolio would instantaneously be at some risk free rate $f(t)$. In other words, the left-hand of (2.2) would have form

$$\begin{aligned} & A_1 dB(t,T_1) + A_2 dB(t,T_2) + A_3 dB(t,T_3) \\ &= d\left[A_1 B(t,T_1) + A_2 B(t,T_2) + A_3 B(t,T_3)\right] \\ &= \left[A_1 B(t,T_1) + A_2 B(t,T_2) + A_3 B(t,T_3)\right] f(t)\,dt. \end{aligned}$$

Substituting back into the left hand determinant in (2.2), yields

$$\begin{vmatrix} 1 & 1 & 1 \\ b^{(1)}(t,T_1) & b^{(1)}(t,T_2) & b^{(1)}(t,T_3) \\ b^{(2)}(t,T_1) & b^{(2)}(t,T_2) & b^{(2)}(t,T_3) \end{vmatrix} f(t) = \begin{vmatrix} f(t,T_1) & f(t,T_2) & f(t,T_3) \\ b^{(1)}(t,T_1) & b^{(1)}(t,T_2) & b^{(1)}(t,T_3) \\ b^{(2)}(t,T_1) & b^{(2)}(t,T_2) & b^{(2)}(t,T_3) \end{vmatrix},$$

or, combining everything on the right hand side

$$\begin{vmatrix} f(t,T_1) - f(t) & f(t,T_2) - f(t) & f(t,T_3) - f(t) \\ b^{(1)}(t,T_1) & b^{(1)}(t,T_2) & b^{(1)}(t,T_3) \\ b^{(2)}(t,T_1) & b^{(2)}(t,T_2) & b^{(2)}(t,T_3) \end{vmatrix} = 0.$$

For this equation to hold for arbitrary T_1, T_2 and T_3, the first row must be some time t dependent linear combination of the second and third rows, that is, there exists a function $\lambda(t) = \left(\lambda^{(2)}(t),\ \lambda^{(2)}(t)\right)^*$ such that for any T

$$
\begin{aligned}
f(t,T) - f(t) &= \lambda^{(1)}(t)\, b^{(1)}(t,T) + \lambda^{(2)}(t)\, b^{(2)}(t,T) \qquad (2.3)\\
\Rightarrow \qquad f(t,T) &= f(t) + \lambda^*(t)\, b(t,T),\\
\Rightarrow \qquad \frac{dB(t,T)}{B(t,T)} &= f(t)\, dt + b^*(t,T)\left[dW(t) + \lambda(t)\, dt\right].
\end{aligned}
$$

A change of measure then yields a model with drift independent of maturity.

In an arbitrage free setting we know that there is a measure equivalent to $\mathbb{P}$ under which asset prices divided by the T-maturing zero coupon bond $B(t,T)$ as numeraire, are martingales, and in Section-1.1 that measure is identified as the forward measure $\mathbb{P}_T$.

A better alternative is to simply define the *forward measure* $\mathbb{P}_T$ (with corresponding BM $W_T(t)$) as the martingale measure induced by $B(t,T)$ as numeraire. Thus for any zero coupon bond $B(t,S)$ with an earlier maturity S (so $t \leq S < T$) we have

$$
\begin{aligned}
d\left(\tfrac{B(t,S)}{B(t,T)}\right) \Big/ \left(\tfrac{B(t,S)}{B(t,T)}\right) &= \left[b(t,S) - b(t,T)\right]^* \left[dW(t) - b(t,T)\, dt\right] \qquad (2.4)\\
&= b(t,S,T)^*\, dW_T(t),
\end{aligned}
$$

where the BM $W_T(t)$ under $\mathbb{P}_T$, and the *bond volatility difference* $b(t,S,T)$ are specified by

$$
dW_T(t) = dW(t) - b(t,T)\, dt \qquad \textit{and} \qquad b(t,S,T) = b(t,S) - b(t,T). \qquad (2.5)
$$

Observe that if S runs through a discrete set of nodes the resulting system of SDEs will clearly have a solution if, for example, the bond differences $b(\cdot,\cdot,T)$ are bounded.

REMARK 2.1 In future please distinguish between the two variable bond volatility functions $b(t,T)$ and the three variable bond volatility difference functions $b(t,T,T_1)$. □

For $T + \delta = T_1$ (replace $S = T$ and $T = T_1$ in the above) the measure change between $\mathbb{P}_T$ and $\mathbb{P}_{T_1}$ will be given by

$$
dW_{T_1}(t) = dW_T(t) + b(t,T,T_1)\, dt. \qquad (2.6)
$$

Substituting zero coupons for the forward contract $F_T(t,T_1)$ using the model free result (1.9), and applying (A.3.3), SDEs for $F_T(t,T_1)$ and its reciprocal are then given by

$$F_T(t,T_1) = \frac{B(t,T_1)}{B(t,T)} \quad \frac{d\left(\frac{B(t,T_1)}{B(t,T)}\right)}{\frac{B(t,T_1)}{B(t,T)}} = [b(t,T_1) - b(t,T)]^* [dW(t) - b(t,T)\,dt],$$

$$\frac{1}{F_T(t,T_1)} = \frac{B(t,T)}{B(t,T_1)} \quad \frac{d\left(\frac{B(t,T)}{B(t,T_1)}\right)}{\frac{B(t,T)}{B(t,T_1)}} = [b(t,T) - b(t,T_1)]^* [dW(t) - b(t,T_1)\,dt].$$

That is, for $t \leq T$, their respective SDEs are

$$\frac{dF_T(t,T_1)}{F_T(t,T_1)} = -b(t,T,T_1)^* \, dW_T(t), \tag{2.7}$$

$$d\left(\frac{1}{F_T(t,T_1)}\right) \Big/ \left(\frac{1}{F_T(t,T_1)}\right) = b(t,T,T_1)^* \, dW_{T_1}(t),$$

showing, as in the HJM framework, that the forward contract $F_T(t,T_1)$ is a martingale under the forward measure $\mathbb{P}_T$ at the start of *its interval* $[T,T_1]$, while its reciprocal is a martingale under the forward measure $\mathbb{P}_{T_1}$ at the end.

REMARK 2.2 The above formulation clearly includes HJM, with the bond volatility $b(t,T)$, and bond volatility difference or forward contract volatility $b(t,T,T_1)$ defined by

$$b(t,T) = -\int_t^T \sigma(t,u)\,du \quad \text{and} \quad b(t,T,T_1) = \int_T^{T_1} \sigma(t,u)\,du$$

respectively. □

2.2 Swaps and swap notation

An interest rate swap mediates risk between two parties by exchanging a sequence of floating Libor rates against a series of fixed coupons.

On the *floating side* we assume fixings and cashflows take place at the *floating side nodes* T_j $(j = 0, 1, 2, ..., n)$, where *Libor* $L(t,T_j)$, which is at a *margin* μ_j to the *cash forward* $K(t,T_j)$

$$L(t,T_j) = K(t,T_j) + \mu_j,$$

is *fixed* to $L(T_j,T_j)$ at node T_j and *paid in arrears* at node T_{j+1}.

The *tenor intervals* $(T_j, T_{j+1}]$ $(j = 0, 1, ..)$ thus defined by the sequence of *floating time nodes* T_j $(j = 0, 1, 2, ..., n)$ are then assumed to be *open on the left* and *closed on the right* because cash usually flows at the end of an interval (after some action during that interval), and tracking cash is after all what is most important in finance! That means thinking of Libor $L(t, T_j)$ being fixed at the end of one interval $(T_{j-1}, T_j]$, and then accruing over the next interval $(T_j, T_{j+1}]$ to be paid at its end.

Intervals are indexed by the unique time node they contain so that $(T_{j-1}, T_j]$ is the jth floating interval for $j = 1, 2, ...$ The index 0 is reserved for the time $T_0 = 0$ unique point to which cashflows are present valued. Indexed quantities should be stored accordingly when programming; for example, $K_j = K(0, T_j)$ should be stored as the jth component of an *initial cash forward vector* K, while $B_{j+1} = B(0, T_{j+1})$ should be stored as the $(j+1)^{th}$ component of an *initial discount vector* B. This scheme turns out to suit C++ in which arrays start at 0, but in MatLab, where arrays start at 1, an extra number is necessary for the zero component.

In practice we are only interested in a finite number of tenor intervals (for example, enough to include all the cashflows of the instruments in some portfolio) so the *terminal node* T_n is assumed to be at a time greater than all other relevant times.

The notation does lead to some contortion (any notation will), for example the jth *coverage* δ_j (period of accrual) of Libor $L(t, T_j)$ is actually the *width* $\Delta_{j+1}$of the $(j+1)^{th}$ interval $(T_j, T_{j+1}]$. We react by thinking of the coverage δ_j as being associated with Libor $L(t, T_j)$ and hence the node T_j, that is

$$\delta_j = \delta(T_j) = \Delta_{j+1} = |(T_j, T_{j+1}]| = T_{j+1} - T_j.$$

But if the width of the interval $(T_j, T_{j+1}]$ was specifically required, for example in an integration routine, we would use Δ_{j+1}. There is little confusion in practice.

Similarly on the *fixed side*, fixings and cashflows are assumed to take place at the *fixed side time nodes* $\overline{T}_i$ $i = (0, 1, 2, ...)$, with *coupon* $\kappa_i = \kappa(\overline{T}_i)$ determined at $\overline{T}_i$, paid at $\overline{T}_{i+1}$ and having the fixed side *coverage*

$$\overline{\delta}_i = \overline{\delta}(\overline{T}_i) = \overline{\Delta}_{i+1} = \left|(\overline{T}_i, \overline{T}_{i+1}]\right| = \overline{T}_{i+1} - \overline{T}_i.$$

Notation for expressions indexed by maturity, like forward measures $\mathbb{P}_T$ (expectation $\mathbf{E}_T$) and BM W_T under $\mathbb{P}_T$, may be simplified in the obvious way when the maturities themselves are indexed, just put

$$\mathbb{P}_{T_j} = \mathbb{P}_j \qquad \mathbf{E}_{T_j} = \mathbf{E}_j \qquad W_{T_j} = W_j.$$

With the above notation the time t value of a *forward starting payer swap* $\text{pSwap}(t)$, in which N floating Libor rates are fixed at $T_{j0}, T_{j1}, ..., T_{jN-1}$ and received in arrears against $M \leq N$ constant coupon κ payments fixed at

$\overline{T}_{i0}, \overline{T}_{i1}, ..., \overline{T}_{iM-1}$ and paid in arrears, is

$$\begin{aligned}
&\text{pSwap}(t) = \text{pSwap}(t, M, N) \\
&= \text{pSwap}\left(t, \kappa, T_{j0}, .., T_{jN}, \overline{T}_{i0}, .., \overline{T}_{iM}, M, N\right) \\
&= \mathbf{E}_0 \left\{ \begin{array}{c} \sum_{j\in\{j0,j1,\ldots,jN-1\}} \dfrac{\beta(t)}{\beta(T_{j+1})} \delta_j L(T_j, T_j) \\ -\sum_{i\in\{i0,i1,\ldots,iM-1\}} \dfrac{\beta(t)}{\beta\left(\overline{T}_{i+1}\right)} \overline{\delta}_i \ \kappa \end{array} \middle| \mathcal{F}_t \right\}, \\
&= \sum_{j=j0}^{jN-1} \delta_j B(t, T_{j+1}) \mathbf{E}_{j+1} \left\{ \left[K(T_j, T_j) + \mu_j \right] \middle| \mathcal{F}_t \right\} - \kappa \sum_{i=i0}^{iM-1} \overline{\delta}_i B\left(t, \overline{T}_{i+1}\right).
\end{aligned}$$

Hence, because $K(t, T_j)$ is a $\mathbb{P}_{T_{j+1}}$-martingale,

$$\begin{aligned}
&\text{pSwap}(t) = \text{pSwap}(t, M, N) \\
&= \sum_{j=j0}^{jN-1} \delta_j B(t, T_{j+1}) L(t, T_j) - \kappa \sum_{i=i0}^{iM-1} \overline{\delta}_i B\left(t, \overline{T}_{i+1}\right), \\
&= B(t, T_{j0}) - B(t, T_{jN}) + \sum_{j=j0}^{jN-1} \delta_j \mu_j B(t, T_{j+1}) - \kappa\, \text{level}\left(t, \overline{T}_{i0}, \overline{T}_{iM}\right),
\end{aligned} \tag{2.8}$$

in terms of the *level function*

$$\begin{aligned}
\text{level}(t) &= \text{level}\left(t, \overline{T}_{i0}, \overline{T}_{iM}\right) = \text{level}\left(t, \overline{T}_{i0}, \overline{T}_{iM}, \overline{\delta}_{i0}, .., \overline{\delta}_{iM-1}\right) \\
&= \sum_{i=i0}^{iM-1} \overline{\delta}_i B\left(t, \overline{T}_{i+1}\right).
\end{aligned} \tag{2.9}$$

Note that *payer* in *payer swap* pSwap (t) means the coupon is paid, in contrast to a *receiver swap* rSwap (t) in which the coupon is received. This use of *payer* and *receiver* is standard and always refers to the coupon.

In these expressions the relevant floating and fixed nodes are respectively

$$\{T_{j0}, T_{j1}, ..., T_{jN}\} \qquad \text{and} \qquad \left\{\overline{T}_{i0}, \overline{T}_{i1}, ..., \overline{T}_{iM}\right\},$$

and if we are dealing concurrently with several swaps we will retain this notation. But if just one swap is under consideration, there is no loss of mathematical clarity and some simplification in modifying the above notation and letting the floating and fixed nodes be respectively

$$\{T_0, T_1, ..., T_N\} \qquad \text{and} \qquad \left\{\overline{T}_0, \overline{T}_1, ..., \overline{T}_M\right\},$$

so that indices are simplified and the payer swap written in the *standard swap form*

$$\begin{aligned}\mathrm{pSwap}(t) &= \sum_{j=0}^{N-1} \delta_j B(t, T_{j+1}) L(t, T_j) - \kappa \sum_{i=0}^{M-1} \overline{\delta}_i B\left(t, \overline{T}_{i+1}\right), \qquad (2.10)\\ &= B(t, T_0) - B(t, T_N) + \sum_{j=0}^{N-1} \delta_j \mu_j B(t, T_{j+1}) - \kappa\, \mathrm{level}\left(t, \overline{T}_0, \overline{T}_M\right).\end{aligned}$$

Because indices begin at 0 (as opposed to say 1) in such formulae, they are easy to program; simply add $j0$ or $i0$ throughout to the respective floating or fixed side indices. Note also, that when we are using this standard form our modification of notation requires that the nodes $\{T_0, T_1, ..., T_N\}$ be regarded as a finite set tailored to a specific swap, with T_0 not necessarily zero.

Continuing this modification of notation, when the meaning is clear we often drop the 0 suffix setting, for example, $T_0 = T$ and $\delta_0 = \delta$. Thus if focused on just one Libor rate or cash forward fixing at T_0 and paying at T_1, we would probably refer to $L(t, T)$ or $K(t, T)$ meaning that the rate is set at $T = T_0$ and paid at $T_1 = T_0 + \delta_0 = T + \delta$. Similarly, in the next equation (2.11), we might set the common maturity T_0 of the forwards to T, and write $F_{T_0}(t, T_{j+1}) = F_T(t, T_{j+1})$.

Introducing the corresponding *forward swap rate*, a weighted average of Libor defined as that value of κ which makes the swap value zero

$$\begin{aligned}\omega(t) = \omega(t, M, N) &= \frac{\sum_{j=0}^{N-1} \delta_j B(t, T_{j+1}) L(t, T_j)}{\sum_{i=0}^{M-1} \overline{\delta}_i B\left(t, \overline{T}_{i+1}\right)} \qquad (2.11)\\ &= \frac{\sum_{j=0}^{N-1} \delta_j F_{T_0}(t, T_{j+1}) L(t, T_j)}{\sum_{i=0}^{M-1} \overline{\delta}_i F_{T_0}\left(t, \overline{T}_{i+1}\right)},\\ &= \frac{B(t, T_0) - B(t, T_N)}{\mathrm{level}\left(t, \overline{T}_0, \overline{T}_M\right)} \quad \text{if} \quad \mu_j = 0 \quad j = 0, .., N-1\end{aligned}$$

the payer swap can be rewritten as the product of the level function and the difference between the swaprate $\omega(t)$ and coupon κ like

$$\begin{aligned}\mathrm{pSwap}(t) &= \left\{ \sum_{i=0}^{M-1} \overline{\delta}_i B\left(t, \overline{T}_{i+1}\right) \right\} [\omega(t) - \kappa]. \qquad (2.12)\\ &= \mathrm{level}(t) [\omega(t) - \kappa].\end{aligned}$$

When the swap starts now at time $t = 0$ with the first fixing at $T_0 = 0$, (2.11) generates the *current swaprate*, a value that is used as the strike for most traded vanilla swaps.

Note that if the margins μ_j are set zero and the level function eliminated between (2.10) and (2.12), the swap can be expressed as

$$\mathrm{pSwap}(t) = \{B(t, T_0) - B(t, T_N)\} \left[1 - \frac{\kappa}{\omega(t)}\right]$$

in terms of just three variables, the swaprate $\omega(t)$ and the zero coupons $B(t,T_0)$ and $B(t,T_N)$ maturing at the start and end of the swap respectively.

To tackle the different fixed and floating schedules and retain clarity of exposition, we often make the further simplifying assumption that each fixed side $\overline{T}_i$ coincides with some floating side T_j, *rolling* at regular multiples r - called the *roll* - of the floating side nodes (for example, $r=1$ for quarterly fixed side coupon, $r=2$ for semiannual coupon and $r=4$ for annual coupon), and in particular

$$\overline{T}_0 = T_0 = 0 \qquad \overline{T}_{i0} = T_{j0} \qquad \overline{T}_{iM} = T_{jN}.$$

That way quantities associated with the fixed side can also be easily indexed to the floating side, and if we know the floating side schedule $\{T_0, T_1, ...\}$ and the roll r, then we can work out the fixed side schedule $\{\overline{T}_0, \overline{T}_1, ...\}$ and fixed side coverages $\{\overline{\delta}_0, \overline{\delta}_1, ...\}$ in terms of the floating nodes and coverages.

To help *map fixed side indices* to corresponding floating side indices we introduce the *index mapping function*

$$\begin{aligned} &J : \{i0, i1, .., iM\} \to \{j0, j1, .., jN\}, \\ &\textit{where} \qquad J(i) = j \quad \Leftrightarrow \quad T_j = \overline{T}_i, \end{aligned} \tag{2.13}$$

which identifies the floating side node T_j corresponding to some fixed side node $\overline{T}_i$.

2.2.1 Forward over several periods

Libors over several periods usually have different margins over the cash forwards, so let the N-period Libor $L^{(N)}(t,T_0)$ over $(T_0, T_N]$ be at margin $\mu^{(N)}$ to the cash forward $K^{(N)}(t,T_0)$. We have

$$L^{(N)}(t,T_0) = K^{(N)}(t,T_0) + \mu^{(N)},$$

$$1 + \delta_0^{(N)} K^{(N)}(t,T_0) = \prod_{j=0}^{N-1} [1 + \delta_j K(t,T_j)] = \frac{1}{F_{T_0}(t,T_N)}, \qquad \delta_0^{(N)} = \sum_{j=0}^{N-1} \delta_j,$$

showing that $K^{(N)}(t,T_0)$ and $L^{(N)}(t,T_0)$ are $\mathbb{P}_{T_N}$-martingales. If $L^{(N)}(t,T_0)$ is fixed at T_0 and exchanged against κ at time T_N (where $T_N = \overline{T}_M$ and $\overline{\delta}_0 = \delta_0^{(N)}$), the present value of the resulting swap is

$$\begin{aligned} \text{pSwap}(t) &= \mathbf{E}_0 \left\{ \frac{\beta(t)}{\beta(T_N)} \delta_0^{(N)} \left(L^{(N)}(T_0,T_0) - \kappa \right) \middle| \mathcal{F}_t \right\}, \\ &= \delta_0^{(N)} B(t,T_N) \left(L^{(N)}(t,T_0) - \kappa \right), \\ &= \sum_{j=0}^{N-1} \delta_j B(t,T_{j+1}) K(t,T_j) - \delta_0^{(N)} B(t,T_N) \left[\kappa - \mu^{(N)} \right]. \end{aligned}$$

In this case the swaprate, which is a martingale under $\mathbb{P}_{T_N}$, is

$$\omega(t) = K^{(N)}(t, T_0) = \frac{\sum_{j=0}^{N-1} \delta_j B(t, T_{j+1}) K(t, T_j)}{\delta_0^{(N)} B(t, T_N)} + \mu^{(N)}.$$

2.2.2 Current time

To locate the *current time* t among the floating side nodes define the *att function* $@(\cdot)$ by

$$@(t) = \inf\left\{\ell \in \mathbb{Z}^+ : t \leq T_\ell\right\}.$$

Thus $@(t) = j$, or simply $@ = j$, means that the present time t is in the jth floating interval $(T_{j-1}, T_j]$, and we can write without confusion

$$T_@ = T_j, \quad T_{@-1} = T_{j-1}, \quad T_{@+k} = T_{j+k}, \dots \text{ etc}$$
$$t \in (T_{@-1},\ T_@] = (T_{j-1},\ T_j] = \mathbb{J}(t) = \mathbb{J}$$

and refer to $\mathbb{J}$ as the *current interval.*

Because neither Libor $L(t, T)$ nor the cash forward $K(t, T)$ *live past their maturity* T, we make the convention that all *maturity dependent functions* $h(t, T)$ *die* at T, that is

$$h(t, T) = h(T, T) \quad \text{for} \quad t > T.$$

So during the current interval $\mathbb{J}$

- functions $h(t, T_@)$ are *alive* until the end of $\mathbb{J}$,
- functions $h(t, T)$ for $T \in \{T_{@+1},\ T_{@+2}, \dots\}$ continue to *live* on subsequent intervals,
- and functions $h(t, T)$ for $T \in \{T_0,\ T_1, \dots,\ T_{@-1}\}$ have *died* at the ends of previous intervals.

These notions will be most useful in developing the spot Libor measure, see Section-5.3.

Chapter 3

Shifted BGM

Shifted (or *displaced diffusion*) BGM generalizes flat BGM by making the combination of a cash forward $K(t,T)$ *plus* a shift $a(T)$ lognormal under $\mathbb{P}_{T+\delta}$ with deterministic volatility $\xi(t,T)$. By varying the shift $a(T)$ the resulting model can be made to move between *flat* BGM $(a(T)=0)$ and Gaussian HJM $(a(T)=\frac{1}{\delta})$. That sort of flexibility is viewed as a plus by those traders who see the market either behaving *normally* or at least something *less than lognormally*. A disadvantage (acceptable in other models) is that rates can go negative.

Shifted BGM, similarly to the CEV model, generates cap or swaption *implied volatility skews* but not smiles. Nevertheless for modest shift values, for example, $0.05 < a(T) < 0.20$, it can *bestfit* a cap or swaption smile quite well. So shifted BGM is a good start for the simple *stochastic volatility* version of BGM, introduced in Chapter-16, in which $\xi(t,T)$ is made stochastic, with the shift part fitting the skew and the stochastic volatility adding curvature to the wings.

The author acknowledges the CEV model, as popularized by Andersen & Andreasen [4], is a viable alternative way of introducing a skew in BGM and has the advantage of retaining positive forwards. But the relative complexity of CEV computations compared to the shifted model, inclines the author to favour the latter for a basic standard skew BGM model.

3.1 Definition of shifted model

In *shifted BGM* the *shifted forward* $H(t,T_j)$ which is the *driver*, cash forward $K(t,T_j)$, Libor and *shift* $a(T_j)=a_j$ $\left(0\le a_j\le\frac{1}{\delta_j}\right)$ are related by

$$H(t,T_j)=K(t,T_j)+a(T_j)=L(t,T_j)-\mu_j+a(T_j), \tag{3.1}$$

where $H(t,T_j)$ is lognormal with deterministic volatility $\xi(t,T)$ under $\mathbb{P}_{T_{j+1}}$

$$\frac{dH(t,T_j)}{H(t,T_j)} = \xi^*(t,T_j)\,dW_{T_{j+1}}(t) \tag{3.2}$$
$$\Rightarrow \quad \frac{H(t,T_j)}{H(0,T_j)} = \mathcal{E}\left\{\int_0^t \xi^*(s,T_j)\,dW_{T_{j+1}}(s)\right\}.$$

Critically, because the shift $a(T_j)$ is time t independent $dH(t,T_j) = dK(t,T_j)$. From the zero coupon bond SDEs (2.1) and (2.4)

$$dH(t,T_j) = H(t,T_j)\,\xi^*(t,T_j)\,dW_{T_{j+1}}(t) = dK(t,T_j) = \frac{1}{\delta_j} d\left(\frac{B(t,T_j)}{B(t,T_{j+1})}\right),$$
$$= \frac{1}{\delta_j}\left(\frac{B(t,T_j)}{B(t,T_{j+1})}\right) b(t,T_jT_{j+1})^*\,dW_{T_{j+1}}(t).$$

so the bond volatility difference and corresponding *measure change* (2.6) are

$$b(t,T_j,T_{j+1}) = \frac{\delta_j H(t,T_j)}{[1+\delta_j K(t,T_j)]}\,\xi(t,T_j) = h_j(t)\,\xi(t,T_j), \tag{3.3}$$
$$\Rightarrow \quad dW_{T_{j+1}}(t) = dW_{T_j}(t) + h_j(t)\,\xi(t,T_j)\,dt,$$

where the jth *drift term* $h_j(t)$ is a martingale under $\mathbb{P}_{T_j}$

$$dh_j(t) = h_j(t)\,[1-h_j(t)]\,\xi^*(t,T_j)\,dW_{T_j}(t). \tag{3.4}$$

Similarly to the flat caplet formula (1.20), in shifted BGM the time t value of a *Black caplet* struck at κ_j, setting at T_j and paying at T_{j+1}, is

$$\mathrm{Cpl}(t) = \mathrm{Cpl}(t,\kappa,T_j) = \mathbf{E}_0\left\{\frac{\delta_j\beta(t)}{\beta(T_{j+1})}\left[L(T_j,T_j)-\kappa_j\right]^+\middle|\,\mathcal{F}_t\right\}, \tag{3.5}$$
$$= \delta_j B(t,T_{j+1})\,\mathbf{E}_{T_{j+1}}\left\{\left[\begin{array}{c} H(t,T_j)\,\mathcal{E}\left(\int_t^{T_j}\xi^*(t,T_j)\,dW_{T_{j+1}}(t)\right) \\ -\left(\kappa_j-\mu_j+a(T_j)\right)\end{array}\right]^+\middle|\,\mathcal{F}_t\right\},$$
$$= \delta_j B(t,T_{j+1})\,\mathbf{B}\left(\left[\begin{array}{c} L(t,T_j)-\mu_j \\ +a(T_j)\end{array}\right], \left[\begin{array}{c}\kappa-\mu_j \\ +a(T_j)\end{array}\right], \sqrt{\int_t^{T_j}|\xi(s,T_j)|^2\,ds}\right).$$

That is, the formula for the price of a caplet in shifted BGM is like that in the flat case, except that the shift is added to both forward and strike.

3.1.1 Several points worth noting

1. The constant elasticity of variance (CEV) model is similar to shifted BGM in that both generate skews but not smiles. Comparing their SDEs in a one-factor framework

$$dL(t,T) = [L(t,T)+a(T)]\,\xi(t,T)\,dW_{T_1}(t),$$
$$dL(t,T) = L^\beta(t,T)\,\gamma(t,T)\,dW_{T_1}(t),$$

and asking that their instantaneous volatilities initially have the same value and slope gives a *rule-of-thumb* relationship between β and $a(T)$

$$a(T) = \frac{(1-\beta)}{\beta} L(0,T).$$

2. For $a(T_j) < \frac{1}{\delta_j}$ clearly $0 < h_j(t) < 1$ and so the bond volatility differences $b(t, T_j, T_{j+1})$ will be positive and bounded. Moreover, from (3.4), the stochastic component of the volatility of $h_j(\cdot)$ lies in $(0,1)$ suggesting the initial value $h_j(0)$ as a possible approximation for stochastic $h_j(t)$.

3. Setting the shift equal to the inverse of the coverage $a(T_j) = \frac{1}{\delta_j}$ (about 4, which is quite large), makes the bond volatility difference (3.3) deterministic

$$b(t, T_j, T_{j+1}) = \frac{\delta_j \left[K(t, T_j) + \frac{1}{\delta_j}\right]}{[1 + \delta_j K(t, T_j)]} \xi(t, T_j) = \xi(t, T_j),$$

and the the model Gaussian. The shift can be negative, but that produces a skew increasing with strike which is not observed in the market. Hence the practical operating range for the shift is $0 \le a_j \le \frac{1}{\delta_j}$; that is, shifted BGM *lives* between flat lognormal and Gaussian skew extremes.

4. The magnitude of the volatility $\xi(\cdot)$ varies with the shift and can be quite different from flat BGM volatility. From (3.3), a *rule of thumb* conversion between shifted models $\langle a_1(T), \xi_1(t,T)\rangle$ and $\langle a_2(T), \xi_2(t,T)\rangle$ is

$$[K(t,T) + a_1(T)]\, \xi_1(t,T) \cong [K(t,T) + a_2(T)]\, \xi_2(t,T).$$

Hence a flat (that is, zero shift) lognormal volatility of 70% converts to 46% at shift $a(T) = 2\%$, and to 0.7% in the Gaussian case at shift $a(T) = 400\%$.

5. An alternative way of setting up shifted BGM is to write it in the *affine form*

$$\begin{aligned} dK(t,T) &= \{\beta(T) K(t,T) + [1-\beta(T)] K(0,T)\} \xi^*(t,T)\, dW_{T_1}(t), \\ &= \left\{K(t,T) + \frac{[1-\beta(T)]}{\beta(T)} K(0,T)\right\} \beta(T) \xi^*(t,T)\, dW_{T_1}(t). \end{aligned}$$

That stabilizes the magnitude of $\xi(t,T)$ as $\beta(T)$ changes because, from the previous *rule of thumb*, for different $\langle \beta(T), \xi(t,T)\rangle$ regimes

$$\begin{aligned} &[\beta_1(T) K(t,T) + [1-\beta_1(T)] K(0,T)]\, \xi_1(t,T) \\ &\cong [\beta_2(T) K(t,T) + [1-\beta_2(T)] K(0,T)]\, \xi_2(t,T), \end{aligned}$$

and the $[\beta(T) K(t,T) + [1-\beta(T)] K(0,T)]$ terms on each side will tend to be similar.

3.2 Backward construction

Using the *bond basics* of Section-2.1 and some of the techniques introduced by Musiela and Rutkowski in [79] and [80] for flat BGM, we can now implement a *backward construction* of shifted BGM.

Temporarily simplify notation by indexing variables by maturity and dropping time t, for example, set

$$B_T = B(t,T) \quad W_T = W_T(t) \quad b_T = b(t,T).$$

For $0 \leq t \leq R < S < T < U$ we will illustrate the method by moving back through the intervals $(T,U]$, $(S,T]$ and $(R,S]$ successively defining the bond volatility differences $[b_T - b_U]$, $[b_S - b_T]$ and $[b_R - b_S]$ in the process.

Our *basic assumption* and starting point is that the deterministic forward volatilities ξ_R, ξ_S and ξ_T are specified exogenously and are bounded, and that K_T is a martingale under $\mathbb{P}_U$. Hence for the three intervals we have

$$\begin{aligned}
\delta_T = |(T,U]| \qquad & \delta_T K_T = \frac{B_T}{B_U} - 1 \qquad H_T = K_T + a_T \\
& dH_T = H_T \xi_T^* dW_U, \\
\delta_S = |(S,T]| \qquad & \delta_S K_S = \frac{B_S}{B_T} - 1 \qquad H_S = K_S + a_S \\
& dH_S = H_S \left[\mu_S dt + \xi_S^* dW_U\right], \\
\delta_R = |(R,S]| \qquad & \delta_R K_R = \frac{B_R}{B_S} - 1 \qquad H_R = K_R + a_R \\
& dH_R = H_R \left[\mu_R dt + \xi_R^* dW_U\right].
\end{aligned}$$

On the *first interval* $(T,U]$, comparison of volatility terms in

$$\delta_T dH_T = \delta_T H_T \xi_T^* dW_U = \delta_T dK_T = d\left(\frac{B_T}{B_U}\right) = \left(\frac{B_T}{B_U}\right)[b_T - b_U]^* dW_U$$

identifies the difference in the bond volatilities

$$[b_T - b_U] = \frac{\delta_T H_T}{1 + \delta_T K_T} \xi_T,$$

which must be bounded (assuming $\delta_T a_T < 1$) because the SDE for H_T has a positive solution, and that defines a change of measure from $\mathbb{P}_U$ to $\mathbb{P}_T$ by

$$dW_T = dW_U - [b_T - b_U]\, dt = dW_U - \frac{\delta_T H_T}{1 + \delta_T K_T} \xi_T dt.$$

On the *second interval* $(S, T]$, in the equation

$$\delta_S dH_S = \delta_S H_S \left[\mu_S dt + \xi_S^* dW_U\right] = \delta_S dK_S = d\left(\frac{B_S}{B_T}\right) = d\left(\frac{B_S}{B_U}\Big/\frac{B_T}{B_U}\right)$$
$$= \left(\frac{B_S}{B_T}\right)\left[b_S - b_T\right]^* \left\{dW_U - \left[b_T - b_U\right] dt\right\},$$

comparison of volatility and drift terms yields

$$\delta_S H_S \xi_S = \left(\frac{B_S}{B_T}\right)\left[b_S - b_T\right], \quad \delta_S H_S \mu_S = -\left(\frac{B_S}{B_T}\right)\left[b_S - b_T\right]^* \left[b_T - b_U\right] \quad \Rightarrow$$
$$\mu_S = -\xi_S^* \left[b_T - b_U\right] \Rightarrow dH_S = H_S \xi_S^* \left[dW_U - \left[b_T - b_U\right] dt\right] = H_S \xi_S^* dW_T.$$

The SDE for H_S has a solution which is strictly positive, which in turn defines the bond volatility difference

$$\left[b_S - b_T\right] = \frac{\delta_S H_S}{1 + \delta_S K_S}\xi_S,$$

and a change of measure from $\mathbb{P}_U$ to $\mathbb{P}_S$ by

$$dW_S = dW_U - \left[b_S - b_U\right] dt = dW_U - \left[(b_S - b_T) + (b_T - b_U)\right] dt$$
$$= dW_U - \left\{\frac{\delta_S H_S}{1 + \delta_S K_S}\xi_S + \frac{\delta_T H_T}{1 + \delta_T K_T}\xi_T\right\} dt.$$

On the *third interval* $(R, S]$ proceed as on the second.

$$\delta_R dH_R = \delta_R H_R \left[\mu_R dt + \xi_R^* dW_U\right]$$
$$= \left(\frac{B_R}{B_S}\right)\left[b_R - b_S\right]^* \left\{dW_U - \left[b_S - b_U\right] dt\right\}$$
$$\Rightarrow \qquad dH_R = H_R \xi_R^* \left[dW_U - \left[b_S - b_U\right] dt\right] = H_R \xi_R^* dW_S$$
$$\Rightarrow \qquad \left[b_R - b_S\right] = \frac{\delta_R H_R}{1 + \delta_R K_R}\xi_R.$$

As before, the SDE for H_R has a strictly positive solution which defines $[b_R - b_S]$ and the next measure change from $\mathbb{P}_U$ to $\mathbb{P}_R$ by

$$dW_R = dW_U - \left[b_R - b_U\right] dt,$$
$$= dW_U - \left\{\frac{\delta_S H_S}{1 + \delta_S K_S}\xi_S + \frac{\delta_T H_T}{1 + \delta_T K_T}\xi_T + \frac{\delta_R H_R}{1 + \delta_R K_R}\xi_R\right\} dt.$$

The measures $\mathbb{P}_R$, $\mathbb{P}_S$ and $\mathbb{P}_T$ thus defined are clearly the forward measures at R, S and T. For example, for any bond B_Q $(Q < T)$ the ratio $\left(\frac{B_Q}{B_T}\right)$ must be a $\mathbb{P}_T$-martingale because

$$d\left(\frac{B_Q}{B_T}\right) = d\left(\frac{B_Q}{B_U}\Big/\frac{B_T}{B_U}\right) = \left(\frac{B_Q}{B_T}\right)\sqrt{V}\left[b_Q - b_T\right]^* dW_T.$$

In general, starting with the *terminal measure* $\mathbb{P}_n$ at the terminal node T_n and working backwards in the above fashion *constructs* the system of equations (1.21) with each forward $K(t, T_j)$ an exponential martingale under the corresponding forward measure $\mathbb{P}_{j+1}$.

REMARK 3.1 The system $\{K(\cdot, T_j) : 0, 1, .., n-1\}$ we have constructed is finite and discrete with variable coverages (an essential feature to fit holidays and business conventions). To make the model operational, however, we often need $K(t, T)$ for maturities T strictly within the tenor intervals; that requires intelligent interpolation methods and is the topic of a later chapter. □

Chapter 4

Swaprate Dynamics

This chapter is devoted to analyzing swaprate dynamics, finding suitable deterministic approximations to swaprate volatilities, and deriving corresponding swaption formulae.

In the shifted-BGM model constructed in the last chapter the forwards plus time independent shifts were dynamically lognormal under the appropriate forward measures. We now show, following [31] (and others - it's the sort of result many implementers have probably obtained independently) that swaprates in this framework exhibit the same sort of behavior as forwards in that swaprates plus shifts that are nearly time independent are dynamically almost lognormal under new measures, equivalent to the forward measures, that we call *swaprate measures.*

Thus in the first Section-4.1 we separate the swaprate into its shift and stochastic parts, and then in Section-4.2 and Section-4.3, analyze each separately and justify our approximations.

A consequence obtained in Section-4.4, is that in shifted-BGM swaptions can be priced accurately with Black type formulae very similar to the ones (3.5) used to price caplets. This is a standard outcome in the BGM framework - whatever method prices caplets, also usually fairly accurately prices swaptions after some necessary adjustments and approximations. For example, see Chapter-16 on the stochastic volatility version of BGM for a similar outcome, or [9] and [95] for generic methods using *Markovian projection.*

When the shift in BGM is zero the swaprate SDEs (4.13) permit, see Section-4.5, the easy derivation of Jamshidian's [66] swaprate model in which the swaprates of coterminal swaps can be made jointly lognormal under a collection of appropriate swaprate measures (rather similar to the forwards under the collection of forward measures in BGM). But more than that, it's possible to construct many other market models in which the swaprates of any set of swaps with a strictly increasing total tenor structure, which may include forwards, can be made jointly lognormal under appropriate measures.

When the shift is non-zero, however, further work is needed to properly construct a Jamshidian type swaprate model that behaves like the forwards in shifted BGM with an exact closed formula for swaptions corresponding to (3.5) for caplets.

Among all the market models, the relative algebraic simplicity of shifted-BGM combined with its ability to handle swap dynamics and calibrate to both caps and swaptions, makes it a *central interest rate model.*

4.1 Splitting the swaprate

From (2.11) and (3.1), and letting $T = T_0 = \overline{T}_0$ the swaprate in shifted BGM is

$$\omega(t) = \frac{\sum_{j=0}^{N-1} \delta_j F_T(t, T_{j+1}) H(t, T_j)}{\sum_{i=0}^{M-1} \overline{\delta}_i F_T(t, \overline{T}_{i+1})} - \frac{\sum_{j=0}^{N-1} \delta_j F_T(t, T_{j+1}) \left[a(T_j) - \mu_j\right]}{\sum_{i=0}^{M-1} \overline{\delta}_i F_T(t, \overline{T}_{i+1})}, \tag{4.1}$$

and we will show that its *stochastic part* $\omega 1(t)$ is almost lognormal

$$\omega 1(t) = \frac{\sum_{j=0}^{N-1} \delta_j F_T(t, T_{j+1}) H(t, T_j)}{\sum_{i=0}^{M-1} \overline{\delta}_i F_T(t, \overline{T}_{i+1})} = \sum_{j=0}^{N-1} u_j(t) H(t, T_j) = \sum_{j=0}^{N-1} v_j(t), \tag{4.2}$$

while what we will call its *shift part* $\omega 2(t)$ is almost constant

$$\omega 2(t) = \frac{\sum_{j=0}^{N-1} \delta_j F_T(t, T_{j+1}) \left[a(T_j) - \mu_j\right]}{\sum_{i=0}^{M-1} \overline{\delta}_i F_T(t, \overline{T}_{i+1})} = \sum_{j=0}^{N-1} u_j(t) \left[a(T_j) - \mu_j\right]. \tag{4.3}$$

The following lemma helps analyze components of the decomposition.

LEMMA 4.1
If $Y(t)$ is an Ito process and $X_0, .., X_{M-1}$ are exponential martingales under some reference measure $\mathbb{P}$

$$\frac{dY(t)}{Y(t)} = \mu(t)\,dt + \xi^*(t)\,dW(t), \quad \frac{dX_i(t)}{X_i(t)} = \sigma_i^*(t)\,dW(t) \quad (i = 0, .., M-1),$$

and $\widetilde{\mathbb{P}}$ is measure equivalent to $\mathbb{P}$ defined by

$$d\widetilde{W}(t) = dW(t) - \frac{\sum_{i=0}^{M-1} X_i(t)\,\sigma_i(t)}{\sum_{i=0}^{M-1} X_i(t)} dt, \qquad \text{then} \tag{4.4}$$

$$\frac{d\left(\frac{Y(t)}{\sum_{i=0}^{M-1} X_i(t)}\right)}{\left(\frac{Y(t)}{\sum_{i=0}^{M-1} X_i(t)}\right)} = \mu(t)\,dt + \left[\xi(t) - \frac{\sum_{i=0}^{M-1} X_i(t)\,\sigma_i(t)}{\sum_{i=0}^{M-1} X_i(t)}\right]^* d\widetilde{W}(t). \quad (4.5)$$

Hence $\frac{1}{\sum X_i}$ is a $\widetilde{\mathbb{P}}$-martingale, and so is $\frac{Y}{\sum X_i}$ when Y is a $\mathbb{P}$-martingale.

PROOF Bearing in mind that each $X_i(t)$ is strictly positive, an SDE for $\sum_{i=0}^{M-1} X_i(t)$ is

$$\frac{d\sum_{i=0}^{M-1} X_i(t)}{\sum_{i=0}^{M-1} X_i(t)} = \frac{\sum_{i=0}^{M-1} X_i(t)\,\sigma_i^*(t)}{\sum_{i=0}^{M-1} X_i(t)} dW(t).$$

The result follows on applying (A.3.3) to $Y(t)$ and $\sum_{i=0}^{M-1} X_i(t)$

$$\frac{d\left(\frac{Y(t)}{\sum_{i=0}^{M-1} X_i(t)}\right)}{\left(\frac{Y(t)}{\sum_{i=0}^{M-1} X_i(t)}\right)} = \mu(t)\,dt + \begin{array}{l}\left[\xi(t) - \frac{\sum_{i=0}^{M-1} X_i(t)\sigma_i(t)}{\sum_{i=0}^{M-1} X_i(t)}\right]^* \\ \times\left[dW(t) - \frac{\sum_{i=0}^{M-1} X_i(t)\sigma_i(t)}{\sum_{i=0}^{M-1} X_i(t)} dt\right]\end{array}.$$

Note that $\widetilde{\mathbb{P}}$ is determined solely by the denominator $\sum_{i=0}^{M-1} X_i(t)$. □

4.2 The shift part

Recalling the weights $u_j(t)$ arise as the coefficients of $H(t, T_j)$ and $a(T_j)$ in the formulae (4.2) and (4.3), and introducing $\overline{u}_i(t)$ and $\overline{f}(t)$, for $j = 0, .., N-1$ and $i = 0, .., M-1$ we have

$$\overline{f}(t) = \frac{1}{\sum_{i=0}^{M-1} \overline{\delta}_i F_T\left(t, \overline{T}_{i+1}\right)}, \quad (4.6)$$

$$u_j(t) = \delta_j F_T(t, T_{j+1})\ \overline{f}(t), \qquad \overline{u}_i(t) = \overline{\delta}_i F_T\left(t, \overline{T}_{i+1}\right)\ \overline{f}(t).$$

In Lemma 4.1 set $X_i(t) = \overline{\delta}_i F_T\left(t, \overline{T}_{i+1}\right)$, $Y(t) = \delta_j F_T(t, T_{j+1})$ with SDEs

$$\frac{d\left[\delta_j F_T(t, T_{j+1})\right]}{\left[\delta_j F_T(t, T_{j+1})\right]} = -b(t, T, T_{j+1})^*\,dW_T(t),$$

$$\frac{d\left[\overline{\delta}_i F_T\left(t, \overline{T}_{i+1}\right)\right]}{\left[\overline{\delta}_i F_T\left(t, \overline{T}_{i+1}\right)\right]} = -b\left(t, T, \overline{T}_{i+1}\right)^* dW_T(t).$$

If the corresponding $\widetilde{\mathbb{P}}$-measure is defined be the *swaprate measure* $\widetilde{\mathbb{P}}_T$ equivalent to $\mathbb{P}_T$ induced by the BM $\widetilde{W}_T(t)$ where

$$d\widetilde{W}_T(t) = dW_T(t) + \sum_{i=0}^{M-1} \overline{u}_i(t)\, b\left(t, T, \overline{T}_{i+1}\right) dt \qquad \textit{then} \tag{4.7}$$

$$\begin{aligned}
\frac{d\overline{f}(t)}{\overline{f}(t)} &= \sum_{i=0}^{M-1} \overline{u}_i(t)\, b^*\left(t, T, \overline{T}_{i+1}\right) d\widetilde{W}_T(t), \\
\frac{du_j(t)}{u_j(t)} &= \left[-b\left(t, T, T_{j+1}\right) + \sum_{i=0}^{M-1} \overline{u}_i(t)\, b\left(t, T, \overline{T}_{i+1}\right)\right]^* d\widetilde{W}_T(t), \\
\frac{d\overline{u}_i(t)}{\overline{u}_i(t)} &= \left[-b\left(t, T, \overline{T}_{i+1}\right) + \sum_{i=0}^{M-1} \overline{u}_i(t)\, b\left(t, T, \overline{T}_{i+1}\right)\right]^* d\widetilde{W}_T(t),
\end{aligned}$$

The best intuition for the swaprate measure $\widetilde{\mathbb{P}}$ is that it is a sort of *average* of the M forward measures $\mathbb{P}_i$ $(i = 1, ..M)$ over the tenor of the underlying forward swap in the sense that

$$d\widetilde{W}_T(t) = \sum_{i=0}^{M-1} \overline{u}_i(t)\, dW_{\overline{T}_{i+1}}(t)$$

(get this result by substituting $dW_{\overline{T}_{i+1}}(t) = dW_T(t) + b\left(t, T, \overline{T}_{i+1}\right) dt$ in (4.7) and using the fact that the $\overline{u}_i(t)$ are weights summing to 1).

Both $u_j(t)$ and $\overline{u}_i(t)$ are $\widetilde{\mathbb{P}}_T$-martingales with small volatilities because the terms they contain will tend to cancel. Their initial values $u_j(0)$ and $\overline{u}_i(0)$ ought therefore to be good approximations to subsequent values. Hence the following *approximation* to the *shift part* (4.3) of the shifted BGM swaprate.

$$\begin{aligned}
\omega 2(t) \cong \omega 2(0) &= \frac{\sum_{j=0}^{N-1} \delta_j F_T\left(0, T_{j+1}\right)\left[a\left(T_j\right) - \mu_j\right]}{\sum_{i=0}^{M-1} \overline{\delta}_i F_T\left(0, \overline{T}_{i+1}\right)} \\
&= \sum_{j=0}^{N-1} u_j(0)\left[a\left(T_j\right) - \mu_j\right].
\end{aligned} \tag{4.8}$$

The drift term appearing in the measure change (4.7), can be rewritten by using (3.3) and changing the order of summation

$$\begin{aligned}
\sum_{i=0}^{M-1} \overline{u}_i(t)\, b\left(t, T, \overline{T}_{i+1}\right) &= \sum_{i=0}^{M-1} \overline{u}_i(t) \sum_{\ell=0}^{J(i+1)-1} h_\ell(t)\, \xi\left(t, T_\ell\right) \\
= \sum_{j=0}^{N-1} \vec{u}_j(t) \sum_{\ell=0}^{j} h_\ell(t)\, \xi\left(t, T_\ell\right) &= \sum_{j=0}^{N-1} \left\{ h_j(t) \sum_{\ell=j}^{N-1} \vec{u}_\ell(t) \right\} \xi\left(t, T_j\right)
\end{aligned}$$

where the vector $\overrightarrow{u}(t)$ is simply the vector $\overline{u}(t)$ with zeros added so as to be able to sum over the floating side j rather than the fixed side i. Specifically

$$\overrightarrow{u}_j(t) = \begin{cases} \overline{u}_i(t) & \text{when} \quad j = J(i+1) - 1 \quad (i = 0, ..., M-1) \\ 0 & \text{when} \quad j \notin \{J(i+1) - 1 : i = 0, 1, ..., M-1\} \end{cases}, \tag{4.9}$$

that is, all the $\overrightarrow{u}_j(t)$ are 0 except for the $\overrightarrow{u}_{J(i+1)-1}(t) = \overline{u}_i(t)$.

4.3 The stochastic part

The second formula concerns the jth term in the stochastic part of the swaprate (4.2)

$$v_j(t) = \frac{\delta_j F_T(t, T_{j+1}) H(t, T_j)}{\sum_{i=0}^{M-1} \overline{\delta}_i F_T\left(t, \overline{T}_{i+1}\right)} \quad j = 0, .., N-1, \tag{4.10}$$

An SDE for the top comes from (3.2), (2.7) and (A.3.3)

$$\frac{dH(t, T_j)}{H(t, T_j)} = \xi^*(t, T_j)\, dW_{T_{j+1}}(t) = \xi^*(t, T_j)\left[b(t, T, T_{j+1})\, dt + dW_T(t)\right]$$

$$\textit{and} \qquad \frac{dF_T(t, T_{j+1})}{F_T(t, T_{j+1})} = -b(t, T, T_{j+1})^*\, dW_T(t) \qquad \Rightarrow$$

$$\frac{d\left[\delta_j F_T(t, T_{j+1}) H(t, T_j)\right]}{\left[\delta_j F_T(t, T_{j+1}) H(t, T_j)\right]} = \{\xi(t, T_j) - b(t, T, T_{j+1})\}^*\, dW_T(t). \tag{4.11}$$

To apply Lemma (4.1) set $Y(t) = \delta_j F_T(t, T_{j+1}) H(t, T_j)$, and then

$$\frac{dv_j(t)}{v_j(t)} = \begin{bmatrix} \xi(t, T_j) - b(t, T, T_{j+1}) \\ +\sum_{i=0}^{M-1} \overline{u}_i(t)\, b\left(t, T, \overline{T}_{i+1}\right) \end{bmatrix}^* d\widetilde{W}_T(t), \tag{4.12}$$

establishing that $v_j(t)$ is an exponential martingale with approximate volatility $\xi(t, T_j)$.

From this equation (4.12) and (4.2), and introducing the weights

$$w_j(t) = \frac{\delta_j F_T(t, T_{j+1}) H(t, T_j)}{\sum_{j=0}^{N-1} \delta_j F_T(t, T_{j+1}) H(t, T_j)} \quad j = 0, .., N-1 \quad \Rightarrow \quad \sum_{j=0}^{N-1} w_j(t) = 1,$$

an exact (no approximations involved) SDE for the stochastic part $\omega 1(t)$ of the swaprate is therefore

$$\frac{d\omega 1(t)}{\omega 1(t)} = \frac{\sum_{j=0}^{N-1} dv_j(t)}{\sum_{j=0}^{N-1} v_j(t)} = \sigma^*(t)\, d\widetilde{W}_T(t) \quad \textit{where} \tag{4.13}$$

$$\sigma(t) = \sum_{j=0}^{N-1} w_j(t) \left[\xi(t, T_j) - b(t, T, T_{j+1}) + \sum_{i=0}^{M-1} \overline{u}_i(t)\, b\left(t, T, \overline{T}_{i+1}\right)\right].$$

Substituting for the bond volatility differences $b(\cdot)$ from (3.3) and changing the order of summation using the index mapping function J (2.13) and the $\overrightarrow{u}_j(\cdot)$-vector (4.9), the *swaprate volatility* $\sigma(t)$ can be expressed as a linear combination of the forward volatilities $\xi(t, T_j)$ as follows

$$\begin{aligned}
\sigma(t) &= \sum_{j=0}^{N-1} w_j(t) \left[\xi(t, T_j) - \sum_{\ell=0}^{j} h_\ell(t) \xi(t, T_\ell) + \sum_{i=0}^{M-1} \overline{u}_i(t) \sum_{\ell=0}^{J(i)} h_\ell(t) \xi(t, T_\ell) \right] \\
&= \sum_{j=0}^{N-1} w_j(t) \xi(t, T_j) - \sum_{\ell=0}^{N-1} \sum_{j=\ell}^{N-1} w_j(t) h_\ell(t) \xi(t, T_\ell) \\
&\quad + \left(\sum_{j=0}^{N-1} w_j(t) \right) \sum_{k=0}^{N-1} \overrightarrow{u}_k(t) \sum_{\ell=0}^{k} h_\ell(t) \xi(t, T_\ell), \\
&= \sum_{j=0}^{N-1} \left\{ w_j(t) - h_j(t) \sum_{\ell=j}^{N-1} \left(w_\ell(t) - \overrightarrow{u}_\ell(t) \right) \right\} \xi(t, T_j).
\end{aligned}$$

Within this expression for $\sigma(t)$ the weights $w_j(t)$ dominate. Further, using (4.11) and Lemma 4.1, SDEs for the weights $w_j(t)$ are

$$\begin{aligned}
\frac{dw_j(t)}{w_j(t)} &= \begin{bmatrix} \xi(t, T_j) - b(t, T, T_{j+1}) \\ -\sum_{j=0}^{N-1} w_j(t) \{\xi(t, T_j) - b(t, T, T_{j+1})\} \end{bmatrix}^* d\widetilde{W}_T^\bullet(t) \\
d\widetilde{W}_T^\bullet(t) &= dW_T(t) - \sum_{j=0}^{N-1} w_j(t) \{\xi(t, T_j) - b(t, T, T_{j+1})\} \, dt,
\end{aligned}$$

showing they are low variance martingales under a measure $\widetilde{\mathbb{P}}_T^\bullet$ induced by $\widetilde{W}_T^\bullet(t)$ and may be approximated by their initial values $w_j(t) \cong w_j(0)$.

That permits a deterministic approximation for the swaprate volatility

$$\sigma(t) \cong \sum_{j=0}^{N-1} w_j(0) \xi(t, T_j)$$

that is suitable, for example, for correlation analysis, see Chapter-6. Moreover, recalling that $h_j(t)$ and $\overrightarrow{u}_\ell(t)$ can be approximated by their initial values, it is tempting to use the deterministic approximation

$$\sigma(t) \cong \sum_{j=0}^{N-1} \left\{ w_j(0) - h_j(0) \sum_{\ell=j}^{N-1} \left(w_\ell(0) - \overrightarrow{u}_\ell(0) \right) \right\} \xi(t, T_j).$$

This approximation is accurate and satisfactory in unshifted BGM, but as was pointed out to the author[1], for large shifts and different fixed and floating

[1] My thanks to Marc-Olivier Seguin for this critical bit of information.

schedules (for example, swaps of quarterly Libor against semi-annual coupon), this formula fails and a different approximation is required.

Start by considering how changes in $v_j(t)$ arise. We have

$$v_j(t) = u_j(t) H(t, T_j) = u_j(t) [K(t, T_j) + a(T_j)],$$

where $a(T_j)$ may be (much) larger than $K(t, T_j)$ and is time independent, while the $u_j(t)$ are virtually constant. So the $u_j(t) a(T_j)$ term varies little in contrast to $u_j(t) K(t, T_j)$, but is much larger, suggesting it should be immediately isolated. That leads to the approximation

$$v_j(t) \cong u_j(t) K(t, T_j) + u_j(0) a(T_j) \qquad \Rightarrow$$
$$\frac{dv_j(t)}{v_j(t)} = \frac{d[u_j(t) K(t, T_j)]}{u_j(t) H(t, T_j)} = \frac{d[u_j(t) K(t, T_j)]}{u_j(t) K(t, T_j)} \frac{K(t, T_j)}{H(t, T_j)} \qquad j = 0, .., N-1.$$

Because both $v_j(t)$ and $u_j(t)$ are $\widetilde{\mathbb{P}}_T$-martingales, and

$$d[u_j(t) K(t, T_j)] = dv_j(t) - a(T_j)\ du_j(t),$$

clearly $u_j(t) K(t, T_j)$ must also be a $\widetilde{\mathbb{P}}_T$-martingale. Also, because the $\mathbb{P}_{T_j}$-forward and $\widetilde{\mathbb{P}}$-swaprate measures are equivalent, we can write

$$\frac{dK(t, T_j)}{K(t, T_j)} = \frac{dH(t, T_j)}{H(t, T_j)} \frac{H(t, T_j)}{K(t, T_j)} = (drift)\, dt + \frac{H(t, T_j)}{K(t, T_j)} \xi^*(t, T_j)\, d\widetilde{W}_T(t).$$

Hence, using the SDE (4.7) for $u_j(t)$, the SDE for $u_j(t) K(t, T_j)$ must be

$$\frac{d[u_j(t) K(t, T_j)]}{u_j(t) K(t, T_j)} = \begin{bmatrix} \frac{H(t,T_j)}{K(t,T_j)} \xi(t, T_j) - b(t, T, T_{j+1}) \\ + \sum_{i=0}^{M-1} \overline{u}_i(t)\, b(t, T, \overline{T}_{i+1}) \end{bmatrix}^* d\widetilde{W}_T(t),$$

and it follows that an approximate SDE for $v_j(t)$ is

$$\frac{dv_j(t)}{v_j(t)} \cong \begin{bmatrix} \xi(t, T_j) \\ + \frac{K(t,T_j)}{H(t,T_j)} \begin{pmatrix} -b(t, T, T_{j+1}) \\ + \sum_{i=0}^{M-1} \overline{u}_i(t)\, b(t, T, \overline{T}_{i+1}) \end{pmatrix} \end{bmatrix}^* d\widetilde{W}_T(t). \qquad (4.14)$$

Note that the two SDEs (4.12) and (4.14) for the *kosher* and approximate $v_j(t)$ both coincide in the flat case.

Hence the following approximate SDE for the stochastic part (4.2) of the swaprate

$$\frac{d\omega 1(t)}{\omega 1(t)} = \sigma^*(t)\, d\widetilde{W}_T(t),$$

$$\sigma(t) \cong \sum_{j=0}^{N-1} w_j(t) \left[\xi(t, T_j) + \frac{K(t, T_j)}{H(t, T_j)} \begin{pmatrix} -b(t, T, T_{j+1}) \\ + \sum_{i=0}^{M-1} \overline{u}_i(t)\, b(t, T, \overline{T}_{i+1}) \end{pmatrix} \right].$$

Substituting for the $b(\cdot)$ and changing the order of summation as above, then expresses $\sigma(t)$ as a linear combination of the $\xi(t, T_j)$

$$\sigma(t) \cong \sum_{j=0}^{N-1} \left\{ w_j(t) - h_j(t) \sum_{\ell=j}^{N-1} \left(\begin{array}{c} w_\ell(t) \dfrac{K(t, T_\ell)}{H(t, T_\ell)} \\ -\vec{u}_\ell(t) \sum_{k=0}^{N-1} \frac{K(t,T_k)}{H(t,T_k)} w_k(t) \end{array} \right) \right\} \xi(t, T_j).$$

Putting everything together yields the following lognormal SDE *approximation* to the *stochastic part* (4.2) of the shifted BGM swaprate

$$\frac{d\omega 1(t)}{\omega 1(t)} = \sigma^*(t)\, d\widetilde{W}_T(t), \qquad \sigma(t) = \sum_{j=0}^{N-1} A_j \xi(t, T_j) \tag{4.15}$$

$$A_j = \left\{ w_j - h_j \sum_{\ell=j}^{N-1} \left(w_\ell \frac{K_\ell}{H_\ell} - \vec{u}_\ell \lambda \right) \right\}, \qquad \lambda = \sum_{\ell=0}^{N-1} \frac{K_\ell}{H_\ell} w_\ell$$

$$w_j = w_j(0) \quad h_j = h_j(0) \quad K_\ell = K(0, T_\ell) \quad H_\ell = H(0, T_\ell) \quad \vec{u}_\ell = \vec{u}_\ell(0).$$

4.4 Swaption values

A *payer swaption* maturing at time T $(= T_0 = \overline{T}_0)$ with strike κ is an option to acquire at T a swap with coupon κ. So its time t value is

$$\begin{aligned} \text{pSwpn}(t) &= B(t, T)\, \mathbf{E}_T \left\{ \sum_{i=0}^{M-1} \overline{\delta}_i B(T, \overline{T}_{i+1}) [\omega(T) - \kappa]^+ \middle| \mathcal{F}_t \right\}, \\ &= \left\{ \sum_{i=0}^{M-1} \overline{\delta}_i B(t, \overline{T}_{i+1}) \right\} \widetilde{\mathbf{E}}_T \left\{ [\omega(T) - \kappa]^+ \middle| \mathcal{F}_t \right\}, \end{aligned}$$

changing between forward measure $\mathbb{P}_T$ and swaprate measure $\widetilde{\mathbb{P}}_T$ using Section-5.1 of the next chapter

$$\mathbf{E}_T \{ f(T) | \mathcal{F}_t \} = \left\{ \sum_{i=0}^{M-1} \overline{\delta}_i F_T(t, \overline{T}_{i+1}) \right\} \widetilde{\mathbf{E}}_T \left\{ \frac{f(T)}{\left\{ \sum_{i=0}^{M-1} \overline{\delta}_i F_T(T, \overline{T}_{i+1}) \right\}} \middle| \mathcal{F}_t \right\}.$$

But from (4.1), (4.8) and (4.15) the swaprate

$$\omega(T) = \omega 1(0)\, \mathcal{E} \left\{ \int_0^T \sum_{j=0}^{N-1} A_j \xi^*(t, T_j)\, d\widetilde{W}_T(t) \right\} - \sum_{j=0}^{N-1} u_j(0) \left[a(T_j) - \mu_j \right].$$

Hence the *present value* of a *swaption* is approximately

$$\text{pSwpn}(0) \cong \left\{ \sum_{i=0}^{M-1} \overline{\delta}_i B\left(0, \overline{T}_{i+1}\right) \right\} \mathbf{B}\left\{\omega(0) + \alpha, \kappa + \alpha, \zeta\right\}, \tag{4.16}$$

$$\zeta^2 = \int_0^T \left| \sum_{j=0}^{N-1} A_j \xi(t, T_j) \right|^2 dt = \sum_{j1=0}^{N-1} \sum_{j2=0}^{N-1} A_{j1} A_{j2} \int_0^T \xi^*(t, T_{j1}) \xi(t, T_{j2})\, dt,$$

$$\alpha = \sum_{j=0}^{N-1} u_j(0) \left[a(T_j) - \mu_j\right],$$

$$A_j = w_j - h_j \sum_{\ell=j}^{N-1} \left(w_\ell \frac{K_\ell}{H_\ell} - \overrightarrow{u}_\ell \lambda \right), \quad \lambda = \sum_{\ell=0}^{N-1} \frac{K_\ell}{H_\ell} w_\ell.$$

Reiterating in part, some terminology for parts of formula (4.16) for future use, includes:

1. The *level function*

$$\text{level}(t) = \sum_{i=0}^{M-1} \overline{\delta}_i B\left(t, \overline{T}_{i+1}\right), \qquad \text{level}(0) = \sum_{i=0}^{M-1} \overline{\delta}_i B\left(0, \overline{T}_{i+1}\right),$$

2. The *swaption shift*

$$\alpha(t) = \sum_{j=0}^{N-1} u_j(t) \left[a(T_j) - \mu_j\right] \qquad \alpha = \sum_{j=0}^{N-1} u_j(0) \left[a(T_j) - \mu_j\right],$$

3. The *swaption zeta*

$$\zeta = \zeta(T) = \beta(T, T_N)\sqrt{T} = \sqrt{\int_0^T \left| \sum_{j=0}^{N-1} A_j \xi(t, T_j) \right|^2},$$

in terms of the *swaption implied volatility* $\beta(T, T_N)$.

4.4.1 Multi-period caplets

Further to Section-2.2.1 which was about simple forwards over many intervals, a *multi-period caplet* is simply a swaption with just one coupon payment at its final maturity. From the definition (4.9) of the $\overrightarrow{u}_\ell$ vector, we have

$$\overrightarrow{u}_\ell = 0 \quad \ell = 0, 1, .., N-2 \qquad \textit{and} \qquad \overrightarrow{u}_{N-1} = 1,$$

so substituting in (4.16), the value of a multi-period caplet in shifted BGM is

$$\mathrm{Cpl}^{(N)}(0) \cong \delta_0^{(N)} B\left(t, \overline{T}_M\right) \mathbf{B}\left\{K^{(N)}(t, T_0) + \alpha, \kappa + \alpha, \zeta\right\},$$

$$\alpha = \sum_{j=0}^{N-1} u_j \left[a(T_j) - \mu^{(N)}\right], \quad \zeta^2 = \int_0^T \left|\sum_{j=0}^{N-1} A_j \xi(t, T_j)\right|^2 dt,$$

$$A_j = w_j + h_j \sum_{\ell=0}^{j-1} \frac{K_\ell}{H_\ell} w_\ell.$$

4.5 Swaprate models

For simplicity we will illustrate construction of Jamshidian's swaprate models with two examples, rather than formal proofs; the reader might like to consult either [66] or [80] for more thorough expositions. The first example is coterminal swaptions, the second is a set of swaps with a strictly increasing total tenor structure.

If shifts and margins are zero (that is, $a(T_j) = 0$ and $\mu_j = 0$ for all j), then from (4.13) an accurate SDE for the swaprate $\omega(t)$ is

$$\frac{d\omega(t)}{\omega(t)} = \sigma^*(t)\, d\widetilde{W}_T(t) \qquad \textit{where}$$

$$\sigma(t) = \sum_{j=0}^{N-1} \left\{ w_j(t) - h_j(t) \sum_{\ell=j}^{N-1} \left(w_\ell(t) - \overrightarrow{u}_\ell(t)\right) \right\} \xi(t, T_j)$$

$$= \sum_{j=0}^{N-1} A_j(t)\, \xi(t, T_j),$$

in which, we emphasize, the $A_j(t)$ are stochastic. To obtain BGM we assumed $\xi(t, T)$ was deterministic, but there is nothing to stop us letting it be stochastic (other than the backward construction technique of Section-3.2 requiring a modification that we do not address here). Our modus-operandi will therefore be to choose stochastic $\xi(t, T_j)$ in a way that makes swaprate volatilities become deterministic.

In our two examples we will assume that fixed and floating sides are quarterly, and the coverage is a uniform δ.

Example 4.1
Start with a set of coterminal swaps, the last three of which are

$$\text{pSwap}_{N-1}(t) = \delta\, B(t, T_N)\left[\omega_{N-1}(t) - \kappa\right],$$

$$\text{pSwap}_{N-2}(t) = \left\{\sum_{i=N-2}^{N-1} \delta_j B(t, T_{j+1})\right\}\left[\omega_{N-2}(t) - \kappa\right],$$

$$\text{pSwap}_{N-3}(t) = \left\{\sum_{i=N-3}^{N-1} \delta_j B(t, T_{j+1})\right\}\left[\omega_{N-3}(t) - \kappa\right].$$

Note that *the total tenor structure* is increasing, that is, the first swap tenor is over $(T_{N-1}, T_N]$, the second is over $(T_{N-2}, T_N]$, and the third over $(T_{N-3}, T_N]$. Now make the swaprates for these three forward swaps lognormal as follows:
[1] The swaprate $\omega_{N-1}(t)$ for $\text{pSwap}_{N-1}(t)$ is simply the forward $K(t, T_{N-1})$ which is lognormal under the forward measure $\mathbb{P}_{T_N}$ when $\xi(t, T_{N-1})$ is deterministic.
[2] The swaprate $\omega_{N-2}(t)$ for $\text{pSwap}_{N-2}(t)$ has volatility

$$\sigma_{N-2}(t) = A_{N-2}^{(N-2)}(t)\,\xi(t, T_{N-2}) + A_{N-1}^{(N-2)}(t)\,\xi(t, T_{N-1})$$

in which $\xi(t, T_{N-1})$ is already determined. Make $\omega_{N-2}(t)$ lognormal under its unique swaprate measure $\widetilde{\mathbb{P}}_{N-2}$ by setting $\sigma_{N-2}(t)$ to whatever deterministic value is required (for example, to fit a swaption implied volatility) by setting

$$\xi(t, T_{N-2}) = \frac{\sigma_{N-2}(t) - A_{N-1}^{(N-2)}(t)\,\xi(t, T_{N-1})}{A_{N-2}^{(N-2)}(t)}.$$

[3] The swaprate $\omega_{N-3}(t)$ for $\text{pSwap}_{N-3}(t)$ has volatility

$$\begin{aligned}\sigma_{N-3}(t) = {} & A_{N-3}^{(N-3)}(t)\,\xi(t, T_{N-3}) + A_{N-2}^{(N-3)}(t)\,\xi(t, T_{N-2}) \\ & + A_{N-1}^{(N-3)}(t)\,\xi(t, T_{N-1})\end{aligned}$$

in which $\xi(t, T_{N-1})$ and $\xi(t, T_{N-2})$ are already determined. Make $\omega_{N-3}(t)$ lognormal under its unique swaprate measure $\widetilde{\mathbb{P}}_{N-3}$ by letting $\sigma_{N-2}(t)$ be deterministic and setting

$$\xi(t, T_{N-3}) = \frac{\sigma_{N-3}(t) - A_{N-2}^{(N-3)}(t)\,\xi(t, T_{N-2}) - A_{N-1}^{(N-3)}(t)\,\xi(t, T_{N-1})}{A_{N-3}^{(N-3)}(t)}.$$

Continuing in this way, the swaprates $\omega_{N-1}(t)$, $\omega_{N-2}(t)$, $\omega_{N-3}(t)$,.... of the coterminal swaps can be made jointly lognormal under the corresponding forward and swaprate measures $\mathbb{P}_{T_N}$, $\widetilde{\mathbb{P}}_{N-2}$, $\widetilde{\mathbb{P}}_{N-3}$,... ▯

Example 4.2
We now show how to make a set of two forwards and the swaprates of two swaps jointly lognormal. Start with the swaps

$$\begin{aligned}
\text{pSwap}_1(t) &= \delta B(t,T_2)\left[K(t,T_1)-\kappa\right],\\
\text{pSwap}_2(t) &= \delta B(t,T_3)\left[K(t,T_2)-\kappa\right],\\
\text{pSwap}_{1,3}(t) &= \delta\left\{B(t,T_2)+B(t,T_3)+B(t,T_4)\right\}\left[\omega_{1,3}(t)-\kappa\right],\\
\text{pSwap}_{3,4}(t) &= \delta\left\{B(t,T_4)+B(t,T_5)\right\}\left[\omega_{3,4}(t)-\kappa\right],
\end{aligned}$$

and notice that they have a *strictly increasing total tenor structure*. Proceed as follows:
[1] Make the forwards $K(t,T_1)$ and $K(t,T_2)$ lognormal under the forward measures $\mathbb{P}_{T_2}$ and $\mathbb{P}_{T_3}$ by letting $\xi(t,T_1)$ and $\xi(t,T_2)$ be deterministic.
[2] The swaprate $\omega_{1,3}(t)$ for $\text{pSwap}_{1,3}(t)$ has volatility

$$\sigma_{1,3}(t) = A_1^{(1,3)}(t)\,\xi(t,T_1) + A_2^{(1,3)}(t)\,\xi(t,T_2) + A_3^{(1,3)}(t)\,\xi(t,T_3)$$

in which $\xi(t,T_1)$ and $\xi(t,T_2)$ are already determined. Make $\omega_{1,3}(t)$ lognormal under its unique swaprate measure $\widetilde{\mathbb{P}}_{1,3}$ by letting $\sigma_{1,3}(t)$ be deterministic and setting

$$\xi(t,T_3) = \frac{\sigma_{1,3}(t) - A_1^{(1,3)}(t)\,\xi(t,T_1) - A_2^{(1,3)}(t)\,\xi(t,T_2)}{A_3^{(1,3)}}.$$

[3] The swaprate $\omega_{3,4}(t)$ for $\text{pSwap}_{3,4}(t)$ has volatility

$$\sigma_{3,4}(t) = A_3^{(3,4)}(t)\,\xi(t,T_3) + A_4^{(3,4)}(t)\,\xi(t,T_4)$$

in which $\xi(t,T_3)$ is already determined. Make $\omega_{3,4}(t)$ lognormal under its unique swaprate measure $\widetilde{\mathbb{P}}_{3,4}$ by letting $\sigma_{3,4}(t)$ be deterministic and setting

$$\xi(t,T_3) = \frac{\sigma_{3,4}(t) - A_3^{(3,4)}(t)\,\xi(t,T_3)}{A_4^{(1,3)}}.$$

One can continue in this way so long as for some T_j there is always a new $\xi(t,T_j)$ available to make the volatility of the next swaprate deterministic, that is, so long as the total tenor structure of the underlying swaps is strictly increasing. $\square$

Chapter 5

Properties of Measures

The previous chapters show that BGM, swaprate models and *market models* in general are in many ways *plays on measures*; just find the right measures to make the variables of interest lognormal! In this chapter we try to capture the essentials of the different measures available.

First note from Geman et al [43] that in general measure changes are governed by:

LEMMA 5.1

If $\mathbb{P}$ and $\mathbb{Q}$ are equivalent measures induced by the numeraires M_t and N_t respectively, then for $0 < t \leq T$ and $f(\cdot)$ measurable $\mathcal{F}_T$

$$\mathbf{E}_{\mathbb{Q}}\left\{ f(T) \middle| \mathcal{F}_t \right\} = \mathbf{E}_{\mathbb{P}}\left\{ \frac{N_T/N_t}{M_T/M_t} f(T) \middle| \mathcal{F}_t \right\}.$$

PROOF For $g(\cdot)$ measurable $\mathcal{F}_T$ we have

$$\mathbf{E}_{\mathbb{Q}}\left\{ \frac{g(T)}{N_T} \middle| \mathcal{F}_t \right\} = \frac{g(t)}{N_t}, \qquad \mathbf{E}_{\mathbb{P}}\left\{ \frac{g(T)}{M_T} \middle| \mathcal{F}_t \right\} = \frac{g(t)}{M_t},$$

$$\Rightarrow \quad \mathbf{E}_{\mathbb{Q}}\left\{ \frac{g(T)}{N_T} \middle| \mathcal{F}_t \right\} = \mathbf{E}_{\mathbb{P}}\left\{ \frac{1/N_t}{M_T/M_t} g(T) \middle| \mathcal{F}_t \right\}.$$

Now set $g(T) = f(T)\, N_T$ and the result follows. ∎

In Section-5.1 we discuss changes between forward and swaprate measures as used in the derivation of the swaption formula of Section-4.4. The two most practical measures for pricing either by simulation or *timeslicer* are the *terminal measure* $\mathbb{P}_n$, for which the zero coupon bond $B(t, T_n)$ is numeraire, and the *spot Libor measure* $\mathbb{P}_0$, for which the *pseudo-bank account* consisting of *rolled up zeros* (an analogue of the HJM bank account) is numeraire.

Using the terminal measure for simulation can lead to blowouts in the sample standard deviation, see Example-5.1, something that does not occur when using Spot Libor. But the terminal measure is technically much easier to use than Spot Libor in timeslicers, see Section-10.

5.1 Changes among forward and swaprate measures

Let $0 < t \le T^* \le T < T_1$ with $f(\cdot)$ measurable $\mathcal{F}_{T^*}$. Change between the *forward measures* $\mathbb{P}_T$ and $\mathbb{P}_{T_1}$ is defined by (2.6)

$$
\begin{aligned}
dW_{T_1}(t) &= dW_T(t) + b(t,T,T_1)\,dt \qquad \Rightarrow \\
\mathbf{E}_T\{f(T^*)|\mathcal{F}_t\} &= \mathbf{E}_{T_1}\left\{\mathcal{E}\left\{\int_t^T b^*(s,T,T_1)\,dW_{T_1}(s)\right\} f(T^*)|\mathcal{F}_t\right\}, \\
&= \mathbf{E}_{T_1}\left\{\frac{F_T(t,T_1)}{F_T(T,T_1)} f(T^*)|\mathcal{F}_t\right\} = F_T(t,T_1)\,\mathbf{E}_{T_1}\left\{\frac{f(T^*)}{F_T(T^*,T_1)}|\mathcal{F}_t\right\},
\end{aligned}
$$

using Girsanov's theorem of Section-A.3.5, and nested conditional expections on the reciprocal of the forward which (integrating the SDE (2.7)) is a $\mathbb{P}_{T_1}$-martingale

$$
\frac{1}{F_T(T,T_1)} = \frac{1}{F_T(t,T_1)}\mathcal{E}\left\{\int_t^T b^*(s,T,T_1)\,dW_{T_1}(s)\right\}.
$$

Note that if T and T_1 are adjacent nodes where $T_1 = T + \delta$, then

$$
\mathbf{E}_T\{f(T^*)|\mathcal{F}_t\} = \frac{1}{1+\delta K(t,T)}\mathbf{E}_{T_1}\{f(T^*)[1+\delta K(T^*,T)]|\mathcal{F}_t\}.
$$

From (4.7), the change between the *forward* measure $\mathbb{P}_T$ and the *swaprate* measure $\widetilde{\mathbb{P}}_T$ is defined by

$$
\begin{aligned}
d\widetilde{W}_T(t) &= dW_T(t) + \sum_{i=0}^{M-1} \overline{u}_i(t)\, b\left(t,T,\overline{T}_{i+1}\right) dt \\
\Rightarrow \quad & \mathbf{E}_T\{X(T^*)|\,\mathcal{F}_t\} \\
&= \widetilde{\mathbf{E}}_T\left\{\mathcal{E}\left\{\int_t^T \sum_{i=0}^{M-1} \overline{u}_i(s)\, b^*\left(s,T,\overline{T}_{i+1}\right) d\widetilde{W}_T(s)\right\} X(T^*)\middle|\, \mathcal{F}_t\right\}, \\
&= \widetilde{\mathbf{E}}_T\left\{\frac{\sum_{i=0}^{M-1}\overline{\delta}_i F_T\left(t,\overline{T}_{i+1}\right)}{\sum_{i=0}^{M-1}\overline{\delta}_i F_T\left(T,\overline{T}_{i+1}\right)} X(T^*)\middle|\, \mathcal{F}_t\right\}, \\
&= \left\{\sum_{i=0}^{M-1}\overline{\delta}_i F_T\left(t,\overline{T}_{i+1}\right)\right\}\widetilde{\mathbf{E}}_T\left\{\frac{X(T^*)}{\left\{\sum_{i=0}^{M-1}\overline{\delta}_i F_T\left(T^*,\overline{T}_{i+1}\right)\right\}}\middle|\, \mathcal{F}_t\right\},
\end{aligned}
$$

using Girsanov Section-A.3.5 and integrating (4.6) for $\overline{f}(t)$ (which is a $\widetilde{\mathbb{P}}_T$-

martingale)

$$\overline{f}(t) = \frac{1}{\sum_{i=0}^{M-1} \overline{\delta}_i F_T\left(t, \overline{T}_{i+1}\right)},$$

$$\frac{\overline{f}(T)}{\overline{f}(t)} = \mathcal{E}\left\{\int_t^T \sum_{i=0}^{M-1} \overline{u}_i(s)\, b^*\left(s, T, \overline{T}_{i+1}\right) d\widetilde{W}_T(s)\right\}.$$

It follows from Lemma 5.1 that the *swaprate numeraire* is simply

$$\sum_{i=0}^{M-1} \overline{\delta}_i B\left(t, \overline{T}_{i+1}\right).$$

5.2 Terminal measure

Under the *terminal measure* $\mathbb{P}_n$ located at T_n

$$dW_{j+1}(t) = dW_n(t) - b(t, T_{j+1}, T_n)\, dt = dW_n(t) - \sum_{\ell=j+1}^{n-1} b(t, T_\ell, T_{\ell+1})\, dt,$$

$$= dW_n(t) - \sum_{\ell=j+1}^{n-1} \tfrac{\delta_\ell H(t, T_\ell)}{[1+\delta_\ell K(t, T_\ell)]} \xi(t, T_\ell)\, dt, \qquad \Rightarrow \tag{5.1}$$

$$\frac{dH(t, T_j)}{H(t, T_j)} = -\left[\sum_{\ell=j+1}^{n-1} \tfrac{\delta_\ell H(t, T_\ell)}{[1+\delta_\ell K(t, T_\ell)]} \xi^*(t, T_\ell)\right] \xi(t, T_j)\, dt + \xi^*(t, T_j)\, dW_n(t),$$

$$H(t, T_j) = H(0, T_j)\, \mathcal{E}\left\{\begin{array}{c} -\int_0^t \left[\sum_{\ell=j+1}^{n-1} \frac{\delta_\ell H(s, T_\ell)}{1+\delta_\ell K(s, T_\ell)} \xi^*(s, T_\ell)\right] \xi(s, T_j)\, ds \\ + \int_0^t \xi^*(s, T_j)\, dW_n(s) \end{array}\right\}. \tag{5.2}$$

Because any asset divided by $B(t, T_n)$ as numeraire is a $\mathbb{P}_n$-martingale, the time t ($\leq T^* \leq T < T_n$) value $X(t)$ of a cashflow $X(T^*)$ determined at T^* and made at T, is given by

$$\frac{X(t)}{B(t, T_n)} = \mathbf{E}_n\left\{\left.\frac{X(T^*)}{B(T, T_n)}\right| \mathcal{F}_t\right\}$$

$$= \mathbf{E}_n\left\{\left.\frac{X(T^*)}{F_T(T, T_n)}\right| \mathcal{F}_t\right\} = \mathbf{E}_n\left\{\left.\frac{X(T^*)}{F_T(T^*, T_n)}\right| \mathcal{F}_t\right\}.$$

In the case when the cashflow is at a node $T = T_j$

$$\frac{X(t)}{B(t, T_n)} = \mathbf{E}_n\left\{\left. X(T^*) \prod_{\ell=j}^{n-1} [1 + \delta_\ell K(T_j, T_\ell)]\right| \mathcal{F}_t\right\}. \tag{5.3}$$

Present valuing a cashflow $X(T_j)$ at T_j by simulation using (5.3), requires averaging the term inside the expectation over all trajectories. But the cashflow $X(T_j)$ for each trajectory must be *forward valued* to T_n by multiplying by the corresponding unbounded

$$\prod_{\ell=j}^{n-1}\left[1+\delta_\ell K\left(T_j,T_\ell\right)\right]$$

before averaging, and that can create a problem when occasional trajectories produce very large forward values.

Example 5.1

Suppose we are simulating a caplet exercising at 5-years in flat BGM under the terminal measure $\mathbb{P}_{10}$ at 10 years with constant forward volatility, of $\xi(t,T)=40\%$ and an intial yieldcurve of 5%. With a quarterly tenor structure $T_\ell=.25\,\ell \quad \ell=0,1,2...$ (making 5-years $=T_{20}$ and 10-years $=T_{40}$)

$$\operatorname{cpl}(0)=.25\ B\left(0,T_{40}\right)\ \mathbf{E}_{40}\left\{\left[K\left(T_{20},T_{20}\right)-.05\right]^{+}\prod_{\ell=21}^{39}\left[1+.25K\left(T_{21},T_\ell\right)\right]\right\}.$$

With $100k$ simulations a positive 4-standard deviation value for the BM driver W_5 will occasionally occur because $\left[1-\mathbf{N}(4)\right]10^5>3$, and from (5.2) that produces forwards $K\left(T_{20},T_j\right) \quad j=20,21,..,39$ of the order of 100% with a forward value factor from T_{21} of approximately

$$\left[1+.25\times 1\right]^{19}\cong 70.$$

The result is to blow out the sample standard deviation, which can only be reduced by making further simulations, which increases computation time. The forward value factor is an inherent deficiency with the terminal measure that makes the author prefer to work with the spot Libor measure for simulation. □

5.3 Spot Libor measure

For the *spot Libor measure* $\mathbb{P}_0$ the *numeraire* is the pseudo-bank *account* $M(t)$ started at time $t=0$ with an investment of 1 in the zero coupon maturing at T_1, which is then rolled over at time $t=T_1$ into the zero coupon maturing at T_2, and so on. Hence

$$M(t) = B(t, T_@) \prod_{\ell=0}^{@-1} \frac{1}{B(T_\ell, T_{\ell+1})} = B(t, T_@) \prod_{\ell=0}^{@-1} [1 + \delta_\ell K(T_\ell, T_\ell)] \quad (5.4)$$

$$\Rightarrow \qquad M(T_j) = \prod_{\ell=0}^{j-1} [1 + \delta_\ell K(T_\ell, T_\ell)] \qquad j = 1, 2, ...$$

On the current interval, the active part of $M(t)$ is the zero coupon $B(t, T_@)$ maturing at time $T_@$, in which we invested $\prod_{\ell=0}^{@-1} \frac{1}{B(T_\ell, T_{\ell+1})}$ at time $T_{@-1}$, and so the dynamics of $M(t)$ on the current interval are therefore solely determined by $B(t, T_@)$. Thus if $Z(t, T_j)$ is the zero $B(t, T_j)$ discounted by the numeraire

$$Z(t, T_j) = \frac{B(t, T_j)}{M(t)}, \quad t \leq T_j, \qquad \textit{then} \quad (5.5)$$

$$\frac{dZ(t, T_j)}{Z(t, T_j)} = -[b^*(t, T_@) - b^*(t, T_j)]\{dW(t) - b(t, T_@)\,dt\},$$
$$= -b^*(t, T_@, T_j)\,dW_0(t),$$

and so $Z(t, T_j)$ will be a martingale under the *spot Libor measure* $\mathbb{P}_0$, determined by the Brownian motion $W_0(t)$ where

$$dW_0(t) = dW(t) - b(t, T_@)\,dt, \qquad \textit{making} \quad (5.6)$$
$$dW_{j+1}(t) = dW_0(t) + b(t, T_@, T_{j+1})\,dt.$$

Substituting into the SDE for $H(t, T_j)$, under $\mathbb{P}_0$ for $j = @(t), .., n-1$, the shifted forward $H(t, T_j)$ satisfies

$$\frac{dH(t, T_j)}{H(t, T_j)} = \xi^*(t, T_j)\,dW_{T_{j+1}}(t) = \xi^*(t, T_j)[b(t, T_@, T_{j+1})\,dt + dW_0(t)],$$

$$= \sum_{\ell=@}^{j} b^*(t, T_\ell, T_{\ell+1})\,\xi(t, T_j)\,dt + \xi^*(t, T_j)\,dW_0(t), \quad (5.7)$$

$$= \left\{\sum_{\ell=@}^{j} \frac{\delta_\ell H(t, T_\ell)}{[1 + \delta_\ell K(t, T_\ell)]}\xi^*(t, T_\ell)\right\}\xi(t, T_j)\,dt + \xi^*(t, T_j)\,dW_0(t),$$

$$H(t, T_j) = H(0, T_j)\,\mathcal{E}\left\{\begin{array}{c} \int_0^t \left[\sum_{\ell=@(s)}^{j} \frac{\delta_\ell H(s, T_\ell)}{1 + \delta_\ell K(s, T_\ell)}\xi^*(s, T_\ell)\right]\xi(s, T_j)\,ds \\ + \int_0^t \xi^*(s, T_j)\,dW_0(s) \end{array}\right\}.$$

Because any asset divided by $M(t)$ as numeraire is a $\mathbb{P}_0$-martingale, the time t ($\leq T^* \leq T < T_n$) value $X(t)$ of a cashflow $X(T^*)$ determined at T^* and made at T, is given by

$$\frac{X(t)}{M(t)} = \mathbf{E}_0 \left\{ \frac{X(T^*)}{M(T)} \middle| \mathcal{F}_t \right\} = \mathbf{E}_0 \left\{ \frac{B(T^*,T)\, X(T^*)}{M(T^*)} \right\}.$$

In the case when the cashflow is at the node $T = T_j$, by

$$\frac{X(t)}{M(t)} = \mathbf{E}_0 \left\{ \frac{X(T^*)}{\prod_{\ell=0}^{j-1} [1 + \delta_\ell K(T_\ell, T_\ell)]} \middle| \mathcal{F}_t \right\}.$$

Note that the unbounded factor $\prod_{\ell=0}^{j-1} [1 + \delta_\ell K(T_\ell, T_\ell)] > 1$ is now in the denominator where it cannot cause blowouts in the payoff.

5.3.1 Jumping measure

Working under the spot Libor measure $\mathbb{P}_0$ is equivalent to working under the successive forward measures $\mathbb{P}_1, ..., \mathbb{P}_j, \mathbb{P}_{j+1}, ..., \mathbb{P}_{n-1}$, jumping from $\mathbb{P}_j$ to $\mathbb{P}_{j+1}$ at time T_j.

To see that, suppose $@(t) = j$, that is $T_@ = T_j$, and concentrate on the current interval $t \in (T_{j-1}, T_j] = (T_{@-1}, T_@] = \mathbb{J}$. Relative to the reference measure $\mathbb{P}$, Brownian motions under the spot measure $\mathbb{P}_0$ and the forward measure $\mathbb{P}_{T_j}$ are respectively given by

$$dW_0(t) = dW(t) - b(t, T_@)\, dt, \qquad dW_j(t) = dW(t) - b(t, T_j)\, dt,$$
$$\Rightarrow \qquad dW_0(t) = dW_j(t).$$

Moreover, the shifted forwards $H(t, T_\ell)$ for $\ell \geq @(t)$ under the spot measure $\mathbb{P}_0$ and the forward measure $\mathbb{P}_{T_j}$ are the same

$$\begin{aligned} \frac{dH(t, T_\ell)}{H(t, T_\ell)} &= \xi^*(t, T_\ell) \left[b(t, T_@, T_{\ell+1})\, dt + dW_0(t) \right], \\ &= \xi^*(t, T_\ell) \left[b(t, T_j, T_{\ell+1})\, dt + dW_j(t) \right]. \end{aligned}$$

Thus for $t \in \mathbb{J}$ all defining variables are identical under both $\mathbb{P}_0$ and $\mathbb{P}_j$. For this reason $\mathbb{P}_0$ is sometimes called the *jumping measure* because it always coincides with the forward measure $\mathbb{P}_@$ at the end of the current interval $\mathbb{J}$.

Chapter 6

Historical Correlation and Volatility

If there are no market instruments implying future correlation, backward looking analysis of historical data is the only reasonable source of information. Hence the topic of this chapter is estimating *historical correlation* within a shifted BGM framework. That said, increasingly in some markets like EUR, there is implied correlation information available in the form of options on differences in short-term and long-term swap rates (CMS spread options), and in the future that will probably become the dominant source of correlation information.

Once determined, correlation can either be input directly into the model becoming part of the parameter set, or it can be used as a desirable *target* for optimization routines that *bestfit* cap and swaption prices.

Our historical data will be assumed to consist of day-by-day quarterly (coverage a uniform $\delta = \frac{1}{4}$ year) readings at maturities $j\delta$ $(j = 0, 1, .., N-1)$ off either yield-to-maturity, forward or swaprate curves. To get quarterly readings, one might use any of the standard curves readily available in banks, designed to fit relevant data and produce discount functions for all maturities.

A practical difficulty is that the (relative) *lack of smoothness* in these standard curves tends to create *phantom principal components.* Thus special curves designed to be *super-smooth* and *bestfit* data are required to give the best results for correlation; for example, use a forward curve specified at quarterly intervals and a bipartite objective function with one part designed to bestfit market data and the other part to minimize the second derivative measured as quarterly second differences.

Another practical difficulty is that the natural objects for analyzing correlation in shifted BGM are forward curves, which must be created from combinations of cash, futures, bond and swap curves by interpolation, differencing and bestfitting. Apart from smoothness problems, when a day-by-day *filmshow* of forward curves is run (always a good test of routines producing periodic curves or surfaces), one often observes *flapping of the long end* of the forward curve, where data is relatively sparse and errors compound.

To analyze day-by-day data like yieldcurves with maturities fixed relative to calendar time t, we introduce the concept of *relative maturity* $x = T - t$ and *relative forward*

$$K(t, x) = K(t, T)|_{T=t+x} \qquad K(t, T) = K(t, x)|_{x=T-t}$$

recycling (despite the possibility of confusion) the notation $K(t,\cdot)$ to avoid creating a new range of variables. Our convention will be that *absolute maturities* use capitals and the variables in which they appear, like $K(t,T)$, are also absolute, while relative maturities use small letters and the variables in which they appear, like $K(t,x)$, are relative. Thus expressions like

$$K(t,t+x) \quad or \quad K(t,T-t)$$

will be avoided as being confusing or meaningless. The ℓth maturity on a yieldcurve changing with calendar time t, will be at the relative maturity x_ℓ with reference to the moving root of that instantaneous yieldcurve, and we will be interested in the set of relative maturities $x_\ell = \ell\delta \ (\ell = 0,..,N-1)$.

To statistically analyze historical correlation, the underlying models describing the day-to-day movement of yield curves must necessarily be stationary. So we assume the shift is a constant a and the *r-factor* shifted BGM historical volatility is *homogeneous* and a function of *relative maturity* $x = (T-t)$ only, that is

$$a(T) = a, \qquad \xi(t,T) = \xi(T-t) = \xi(x) = \left(\xi^{(1)}(x), \xi^{(2)}(x), .., \xi^{(r)}(x)\right)^*.$$

The *correlation function* $c(\cdot)$ defined by

$$c(T-t) = c(x) = \frac{\xi(x)}{\|\xi(x)\|} = \frac{\xi(T-t)}{\|\xi(T-t)\|} \qquad \Rightarrow \qquad \|c(\cdot)\| = 1$$

carries the *correlation* $\rho_{j,k}$ between the shifted relative forwards $H(t,x_j)$ and $H(t,x_k)$ because for $j,k = 0,..,N-1$

$$\rho_{j,k} = \frac{\xi^*(x_j)\,\xi(x_k)}{\|\xi(x_j)\|\,\|\xi(x_k)\|} = c^T(x_j)\,c(x_k). \tag{6.1}$$

The corresponding $N \times N$ *correlation matrix* R for shifted relative forwards at relative maturities $x_\ell = \ell\delta \ (\ell = 0,..,N-1)$ is then

$$R = (\rho_{j,k}) = (c^*(x_j)\,c(x_k)).$$

Note that the relative maturity $x_0 = 0$ corresponds to the present instantaneous forward $K(t,t)$ and so will not figure in option volatility calculations (which are specified as expectations of payoffs involving forwards that have not yet matured). For that reason the first row and column of R are often discarded as irrelevant.

From the time t independent correlation $\rho_{j,k}$ between relative shifted forwards, the *instantaneous correlation* $\rho_{j,k}(t)$ between the absolute forwards $H(t,T_j)$ and $H(t,T_k)$ is

$$\rho_{j,k}(t) = \frac{\xi(t,T_j)^*\,\xi(t,T_k)}{\|\xi(t,T_j)\|\,\|\xi(t,T_k)\|} = \frac{\xi(T_j-t)^*\,\xi(T_k-t)}{\|\xi(T_j-t)\|\,\|\xi(T_k-t)\|}.$$

An important property of the vector volatility function $\xi(t,T)$ is that scaling it by either a deterministic or stochastic scalar quantity $\psi(t,T)$, leaves the instantaneous correlation $\rho_{j,k}(t)$ unchanged because

$$\frac{\psi(t,T_j)\,\xi(t,T_j)^*\,\psi(t,T_k)\,\xi(t,T_k)}{\|\psi(t,T_j)\,\xi(t,T_j)\|\,\|\psi(t,T_k)\,\xi(t,T_k)\|} = \frac{\xi(t,T_j)^*\,\xi(t,T_k)}{\|\xi(t,T_j)\|\,\|\xi(t,T_k)\|} = \rho_{j,k}(t).$$

As we will see in Chapter-7, this flexibility to retain instantaneous correlation while scaling by an arbitrary $\psi(t,T)$ will permit both instantaneous historical correlation and cap and swaption implied volatilities to be jointly fitted. Also, from (3.3),

$$b(t,T_j,T_{j+1}) = \frac{\delta_j H(t,T_j)}{[1+\delta_j K(t,T_j)]}\,\xi(t,T_j) \qquad \Rightarrow$$

$$\rho_{j,k}(t) = \frac{\xi(t,T_j)^*\,\xi(t,T_k)}{\|\xi(t,T_j)\|\,\|\xi(t,T_k)\|} = \frac{b(t,T_j,T_{j+1})^*\,b(t,T_k,T_{k+1})}{\|b(t,T_j,T_{j+1})\|\,\|b(t,T_k,T_{k+1})\|},$$

showing the instantaneous correlation between the forward contracts

$$F_{T_j}(t,T_{j+1}) \qquad and \qquad F_{T_k}(t,T_{k+1})$$

is the same as that between the corresponding shifted forwards $H(t,T_j)$ and $H(t,T_k)$. That raises the possibility that correlation can be estimated in one of the Gaussian, or flat or shifted BGM frameworks, and reasonably used in the others.

Various possible ways include:

- The Gaussian HJM approach which is straightforward, because within it correlation analysis is on yield-to-maturity curves, which are stable, well defined, and do not have to be particularly smooth.

- In contrast, the direct flat BGM approach operates on forward curves, which are essentially differenced yield-to-maturity curves that must be interpolated, are therefore potentially unstable, and which must be smooth to avoid *phantom principle components*.

- Another flat BGM approach is to first estimate swaprate correlation and then back out forward correlation. The point is that in *swap-world* the first 5 principal components are stable (sometimes for years), capture most (around 99.99%) of the R^2, and have flatish tails.

- An approach more consistent with the data is to estimate an *average shift* $a = a(T)$ from option data (see the next Chapter-7), and then find the correlation by adapting the flat BGM techniques.

6.1 Flat and shifted BGM off forwards

Differentiate $K(t,x)$ using the standard SDE

$$dK(t,T) = K(t,T)\,\xi^*(t,T)\,dW_{T_1}(t),$$

and setting $T = t + x$ to formally obtain

$$\begin{aligned}
dK(t,T) &= dK(t,x) - \frac{\partial K(t,x)}{\partial x}\,dt = K(t,x)\,\xi^*(x)\,dW_{T_1}(t), \\
\frac{dK(t,x)}{K(t,x)} &= \frac{1}{K(t,x)}\left\{\begin{array}{c} \frac{\partial K(t,x)}{\partial x} \\ +K(t,x)\,\xi^*(x)\,b(t,T_{@},T_1) \end{array}\right\} dt + \xi^*(x)\,dW_0(t), \\
&= (\mathit{drift})\,dt + \xi^*(x)\,dW_0(t),
\end{aligned} \tag{6.2}$$

as an SDE for $K(t,x)$ under the spot Libor measure $\mathbb{P}_0$.

REMARK 6.1 A rigorous derivation of an equation like (6.2) requires a continuous spectrum of forwards for all maturities as in the framework of the forward BGM construction. But what is really relevant about (6.2) is that the time dependence in the relative maturity $x = T - t$ affects only the drift in the SDE for $K(t,x)$. Moreover, that holds in general; when relative maturities are substituted for absolute maturities in variables like forwards, zero coupon bonds or swaprates, their SDEs change only in their drift. But the actual form of the drift is unimportant because it makes no contribution in the quadratic variation estimators we will use. For similar reasons the measure underlying the SDE, like $\mathbb{P}_0$ in (6.2), is also unimportant. □

A quadratic variation estimator for the *covariance* matrix q of the N forward volatilities $\xi^*(x_j)$ $(j = 0, 1, ...N-1)$ is then

$$q = (q_{j,k}) = (\xi^*(x_j)\,\xi(x_k)) = \left(\frac{1}{t}\int_0^t \frac{d\langle K(\cdot,x_j),K(\cdot,x_k)\rangle(s)}{K(s,x_j)\ K(s,x_k)}\right). \tag{6.3}$$

Assembling the first r (by size of eigenvalue) principal components of q into the $N \times r$ *square root* Ξ of q (so $q = \Xi\,\Xi^T$), the vector volatility function $\xi(x)$ at the discrete points x_j can be recovered from q by setting the transpose of $\xi(x_j)$ equal to the appropriate row in Ξ, that is

$$\xi^T(x_j) = \Xi_{j,\cdot} \quad j = 0, 1, ...N-1.$$

Then, for a continuous $\xi(x)$, simply interpolate between these $\xi(x_j)$.

A similar estimator to (6.3) for the shifted case follows from (3.1)

$$\frac{dK(t,x)}{K(t,x)+a} = (drift)\,dt + \xi^*(x)\,dW_0(t) \tag{6.4}$$
$$\Rightarrow \quad \xi^*(x)\,\xi(y) = \frac{1}{t}\int_0^t \frac{d\langle K(\cdot,x), K(\cdot,y)\rangle(s)}{[K(s,x)+a]\ [K(s,y)+a]},$$

where we have made the shift function $a(T)$ homogeneous, by averaging it into a constant a (the next Chapter-7 shows how that might be done).

For a given set of data, the shifted covariance estimator (6.4) will vary with the shifts. But note that its denominator (which is always positive and almost invariant for large shifts) will tend to just scale the covariance matrix and so not alter correlation, in contrast, the numerator has a more dramatic effect. So approximate this estimator by averaging forwards $K(t,x) \to K(x)$ and shifts $a(T) \to a$ and using

$$\xi^*(x)\,\xi(y) \cong \frac{1}{t}\int_0^t \frac{d\langle K(\cdot,x), K(\cdot,y)\rangle(s)}{[K(x)+a]\ [K(y)+a]}.$$

The resulting covariance matrices vary with shift, but the corresponding correlation matrices are the same. That supports the notion of *one correlation fits all* in shifted BGM. A sensible approach, however, is to use a Gaussian estimator for large shifts, and a lognormal one for small shifts.

6.2 Gaussian HJM off yield-to-maturity

From (1.4) the SDE for the zero coupon $B(t,T)$ under the HJM spot measure $\mathbb{P}_0$ is

$$\frac{dB(t,T)}{B(t,T)} = r(t)\,dt - \int_t^T \sigma^*(t,u)\,du dW_0(t) = r(t)\,dt + b^*(t,T)\,dW_0(t),$$

in which we now assume

$$\sigma(t,T) = \sigma(T-t) \quad \Rightarrow \quad b(t,T) = -\int_t^T \sigma(u-t)\,du = -\int_0^{T-t} \sigma(v)\,dv = b(x).$$

Define the *relative maturity zero coupon bond* $B(t,x)$ by

$$\begin{aligned} B(t,T) &= B(t,x)|_{x=T-t} \qquad \textit{with SDE} \\ dB(t,x) &= dB(t,T) + \frac{\partial B(t,x)}{\partial x}\,dt, \\ &= \left[\frac{\partial B(t,x)}{\partial x} + B(t,x)\,r(t)\right]dt + B(t,x)\,b^*(x)\,dW_0(t), \quad or \\ \frac{dB(t,x)}{B(t,x)} &= (\textit{drift})\,dt + b^*(x)\,dW_0(t). \end{aligned}$$

Hence an estimator for the bond volatility *covariance* is

$$b^*(x)\,b(y) = \frac{1}{t}\int_0^t \frac{d\langle B(\cdot,x), B(\cdot,y)\rangle(s)}{B(s,x)\ B(s,y)}.$$

An alternative estimator uses yields-to-maturity rather than zero coupons

$$\begin{aligned} Y(t,x) &= -\frac{1}{x}\ln P(t,x), \quad \textit{with SDE} \quad dY(t,x) = (\textit{drift})\,dt - \frac{b^*(x)}{x}\,dW_0(t) \\ \Rightarrow \quad & \frac{b^*(x)}{x}\frac{b(y)}{y} = \frac{1}{t}\int_0^t d\langle Y(\cdot,x), Y(\cdot,y)\rangle(s). \end{aligned}$$

Principal component analysis readily yields an $r \times N$ matrix for $x_j = j\delta$

$$\left(b^T(x_j)\right) = \left(b^T(j\delta)\right) = \left(b_{j,\cdot}\right)$$

containing the principal components in the columns, from which $b(x)$ can be interpolated. The shifted forward volatility $\xi(t,T)$ can then be recovered by differencing (recall the remarks of Section-3.1.1)

$$\begin{aligned} \xi(t,T_j) = b(t,T_j,T_{j+1}) = b(t,T_j) - b(t,T_{j+1}) &= b(T_j - t) - b(T_{j+1} - t), \\ \textit{that is} \qquad \xi(x) = b(x) - b(x+\delta) \qquad & \textit{and} \\ c(T-t) = c(x) = \frac{b(x) - b(x+\delta)}{\|b(x) - b(x+\delta)\|}. \end{aligned}$$

6.3 Flat and shifted BGM off swaprates

From (2.11), for $j = 0, 1, \ldots N-1$ define the *relative swaprate* $\omega(t,j)$ of a quarterly swap consisting of $j+1$ rolls fixing at relative maturities $x_\ell = \ell\delta$ $(\ell = 0, 1, .., j)$, to be

$$\omega(t,j) = \frac{\sum_{\ell=0}^{j} B(t,x_{\ell+1})\,K(t,x_\ell)}{\sum_{\ell=0}^{j} B(t,x_{\ell+1})}. \tag{6.5}$$

Using the last yield curve in the historical set, in (4.15) set $A_\ell = w_\ell$ (which is a good enough approximation for correlation work) and differentiate (6.5) using (6.2). The time dependence in the *relative maturity* variable $x_\ell =$ in the bonds adds terms only to drifts, and so under the spot measure $\mathbb{P}_0$ the formal SDE (see Remark-6.1) for the jth swaprate $\omega(t,j)$ must have form

$$\frac{d\omega(t,j)}{\omega(t,j)} = (drift)\,dt + \sigma^*(j)\,dW_0(t)\,, \quad \textit{where approximately} \tag{6.6}$$

$$\sigma(j) = \sum_{\ell=0}^{j} c_{j,\ell}\,\xi(x_\ell)\,, \quad j = 0,1,...,N-1 \quad \textit{in which} \tag{6.7}$$

$$c_{j,\ell} = \begin{cases} \dfrac{B(0,x_{\ell+1})\,K(0,x_\ell)}{\sum_{\ell=0}^{j} B(0,x_{\ell+1})\,K(0,x_\ell)} & \ell = 0,..,j \\ 0 \quad \ell = j+1,..,N-1 \end{cases}$$

Collecting the weights c_j,ℓ into the $N \times N$ lower triangular matrix

$$C = \begin{pmatrix} c_{0,0} & 0 & \cdots & 0 \\ c_{1,0} & c_{1,1} & \cdots & 0 \\ \vdots & \vdots & \ddots & \vdots \\ c_{N-1,0} & c_{N-1,1} & \cdots & c_{N-1,N-1} \end{pmatrix} \tag{6.8}$$

the N equations (6.7) for $j = 0,1,...,N-1$ can be rewritten

$$\Omega = C\,\Xi,$$

in terms of the $N \times r$ matrices Ξ and Ω containing discrete values of $\xi(\cdot)$ and $\sigma(\cdot)$

$$\Xi = (\Xi_{j,\cdot}) = (\xi(x_j))\,, \quad \Omega = (\Omega_{j,\cdot}) = (\sigma(j)) \quad j = 0,..,N-1.$$

Note $c_{0,0} = 1$ while the order of magnitude of $c_{j,\ell}$ is roughly $\frac{1}{j}$, that is, the $c_{j,\ell}$ get smaller by the row. Clearly C is non-singular, so subtracting the restructured $(j-1)^{th}$ and j^{th} equations in (6.7) for $j = 1,..,N-1$

$$\sum_{\ell=0}^{j-1} B(0,x_{\ell+1})\,K(0,x_\ell)\;\xi(x_\ell) = \sigma(j-1)\left\{\sum_{\ell=0}^{j}\left[\begin{array}{c} B(0,x_{\ell+1})\,K(0,x_\ell) \\ -B(0,x_{j+1})\,K(0,x_j) \end{array}\right]\right\},$$

$$\sum_{\ell=0}^{j} B(0,x_{\ell+1})\,K(0,x_\ell)\;\xi(x_\ell) = \sigma(j)\sum_{\ell=0}^{j} B(0,x_{\ell+1})\,K(0,x_\ell)$$

gives the forward volatility $\xi(\cdot)$ in terms of the swaprate volatility $\sigma(\cdot)$

$$\begin{aligned}
\xi(x_j) &= [\sigma(j) - \sigma(j-1)] \frac{\sum_{\ell=0}^{j} B(0, x_{\ell+1}) K(0, x_\ell)}{B(0, x_{j+1}) K(0, x_j)} + \sigma(j-1), \\
&= \left(1 - \frac{1}{c_{j,j}}\right) \sigma(j-1) + \frac{1}{c_{j,j}} \sigma(j), \\
&= \sigma(j) + [\sigma(j) - \sigma(j-1)] \left(\frac{1}{c_{j,j}} - 1\right), \qquad (6.9)
\end{aligned}$$

revealing the inverse of C to be the bi-diagonal matrix

$$C^{-1} = \begin{pmatrix}
1 & 0 & 0 & \cdots & \cdots & \cdots & 0 \\
1 - \frac{1}{c_{1,1}} & \frac{1}{c_{1,1}} & 0 & \ddots & \ddots & \ddots & 0 \\
0 & \ddots & \ddots & 0 & \ddots & \ddots & \vdots \\
\vdots & 0 & 1 - \frac{1}{c_{j-1,j-1}} & \frac{1}{c_{j-1,j-1}} & 0 & \ddots & \vdots \\
\vdots & \ddots & 0 & 1 - \frac{1}{c_{j,j}} & \frac{1}{c_{j,j}} & 0 & \vdots \\
0 & \ddots & \ddots & 0 & \ddots & \ddots & 0 \\
0 & \cdots & \cdots & 0 & 0 & 1 - \frac{1}{c_{N-1,N-1}} & \frac{1}{c_{N-1,N-1}}
\end{pmatrix}. \qquad (6.10)$$

From (6.6), a quadratic variation estimator for the *swaprate covariance* matrix Q for the N swaprate volatilities $\sigma(j)$ is for $j, k = 0, .., N-1$,

$$Q = (Q_{j,k}) = \left(\sigma^T(j)\, \sigma(k)\right) = \left(\frac{1}{\tau} \int_0^\tau \frac{d\langle \omega(\cdot, j), \omega(\cdot, k)\rangle(t)}{\omega(t, j)\, \omega(t, k)}\right), \qquad (6.11)$$

from which $\left(\sigma^T(j)\right)$ can be found by principal component analysis

$$Q = \Omega \Omega^T \qquad \sigma^T(j) = \Omega_{j,\cdot} \quad j = 0, 1, ...N-1$$

That yields the corresponding forward volatility function $\xi(\cdot)$ via

$$(\xi(x_j)) = (\Xi_{j,\cdot}) = \Xi = C^{-1}\Omega,$$

and the corresponding *implied forward covariance* matrix

$$q = (q_{j,k}) = \Xi\Xi^T = C^{-1} Q \left(C^{-1}\right)^T,$$

from which we can compute the correlation function

$$c(T-t) = c(x) = \frac{\xi(x)}{\|\xi(x)\|}.$$

REMARK 6.2 From equation (6.9) the magnitude of $\xi(x_j)$ is roughly

$$\xi(x_j) \cong \sigma(j) + (j-1)[\sigma(j) - \sigma(j-1)].$$

So if in swap-world a principal component appearing in a column of $\left(\sigma^T(j)\right)$ becomes steep in the tail (where j approaches N), then the tail of the corresponding column of $\left(\xi^T(x_\ell)\right)$ will be even steeper, leading to gross distortion of the correlation matrix in forward-world. ▯

Using (4.8), (4.15) and (4.16), a similar estimator to (6.11) for the *swaprate covariance* matrix in the shifted case comes from

$$\frac{d\omega(t,j)}{\omega(t,j)+\alpha} = (drift)\, dt + \sigma^*(j)\, dW_0(t)$$
$$\Rightarrow \quad \sigma^T(j)\,\sigma(k) = \frac{1}{\tau}\int_0^\tau \frac{d\langle\omega(\cdot,j),\omega(\cdot,k)\rangle(t)}{[\omega(t,j)+\alpha]\,[\omega(t,k)+\alpha]},$$

where the maturity dependent swaprate shift has been averaged to a constant α (see Chapter-7 for how that might be done).

Chapter 7

Calibration Techniques

Most banks continuously monitor interest rate market data and after stripping, interpolating, and massaging it, present the processed data as *market objects* (in the programming sense) ready to be used in pricing algorithms. Typically the *object* will incorporate yieldcurve and volatility information for a given currency and particular sector (like Treasuries or swaps or municipals in USD).

So in this chapter we assume volatility information is available in the form of *interpolated swaption* and *stripped caplet implied volatilities* entered in a quarterly *implied volatility matrix* with *exercise times* descending row-by-row and *tenors* moving left to right along the columns. If data is available away-from-the-money it is entered into identical but separate volatility matrices referenced either by *absolute strike* or *relative delta strike*, the whole making up an implied *volatility cube.*

Example 7.1

In a 40×40 implied volatility matrix position $(1,1)$ is the top left-hand corner, the volatility of the caplet exercising at $9\frac{3}{4}yrs$ is in the $(39,1)$ position, and the volatility of the swaption exercising at $5yrs$ into a $5yr$ swap (that is, the swap's *tenor* is $5yrs$) is in the $(20,20)$ position. Because

$$(maturity) = (exercise\ time) + (tenor),$$

both are included among the volatilities of the set of 39 *coterminal* swaptions *maturing* at $10yrs$, which appear along the *diagonal* $\{(i,j) : i+j = 40\}$. □

Some entries in the cube are direct liquid market quotes which a *good calibration* would return exactly. Others are *interpolated* between the liquid quotes in some fashion, which can potentially introduce arbitrage opportunities (the author is unaware of any criteria for ensuring an arbitrage free interpolation, other than it be generated by an arbitrage free interest rate model first fitted to the liquid data).

In shifted BGM the shift and volatility are *orthogonal* in the sense that the shift $a(T)$ can be found independently, as in Section-7.1 below, leaving the volatility $\xi(t,T)$ to be fitted in exactly the same way as in flat BGM (though magnitudes will differ with shift). But note that while the volatility $\xi(t,T)$

with two parameters can potentially fit a whole volatility matrix, the shift $a(T)$ with only one parameter is more restricted.

Many of the volatility functions $\xi(t,T)$ we use are of the form

$$\xi(t,T) = \chi(t)\ \phi(T)\ \psi(T-t)\ c(T-t),$$

in which $\chi(\cdot)$, $\phi(\cdot)$ and $\psi(\cdot)$ are scalars and $c(\cdot)$ is the vector *historical correlation* function as constructed in Chapter-6. Such functions return the correct *instantaneous correlation*, can *exactly fit caps* and a *diagonal of coterminal swaptions* (allowing vegas with respect to them to be computed), can be made reasonably homogeneous (stationary), and in addition can be made to approximately fit a selection of other swaptions (boosting confidence in all-round pricing).

Another approach, which essentially depends on the power of optimisers like the NAG [81] or IMSL [64] ones, is to *bestfit* a selection of liquid caps and swaptions in a *generic calibration*. Usually the motive is to generate *indicative prices* to help increase comfort levels with a shortrate model that is actually used for marking-to-market, risk management and hedging. The BGM model incorporates more volatility and correlation information than the shortrate model, whose comparative advantage is being Markov in a few variables and so fast.

Among the *bestfit* genre, *Pedersen's method* [85] in which

$$\xi(t,T) = \xi(t,T)|_{T=t+x} = \xi(t,x)$$

is made piecewise constant over quarterly intervals in both t and x directions, works well and is simple to implement. We also describe the *cascade algorithm* in which $\xi(t,T)$ is made piecewise constant over quarterly intervals in both t and T directions, but which then requires smoothing to be made usable (see Brigo and Mercurio [32] for advice on how to do that).

Recall that swaption, but not caplet, implied volatilities depend weakly on correlation. Using *semi-definite programming* and a Pessler type volatility function, we also show how correlation can be freed up to participate in calibration (while staying close to a historical target) and permit an exact fit to the whole matrix. Handling the large covariance matrices involved (120×120 or bigger) with presently available semi-definite programming software is, however, problematical although Raise Partner [100] evidently has software that works.

The next section shows *how to fit the skew* with an appropriate term structure of shift $a(T)$, and at the same time *identifies* the corresponding *zetas* $\zeta(\cdot)$ and *implied volatilities* $\beta(\cdot)$ of caplets and swaptions

$$\zeta^2(T) = \beta^2(T)\ T = \int_0^T |\xi(t,T)|^2\, dt,$$

$$\zeta^2(T_j,T_N) = \beta^2(T_j,T_N)\ T_j = \int_0^{T_j} \left| \sum_{\ell=j}^{N-1} A_\ell^{(j,N)} \xi(t,T_\ell) \right|^2 dt,$$

which must then be fitted with a suitable choice of volatility function $\xi(t,T)$.

7.1 Fitting the skew

Fitting the *caplet skew* and isolating the volatility is straightforward.

Step-1: For fixed T_j suppose we have the implied volatilities $\beta_\ell(T_j)$ for $\ell = 1,2,...$ of a set of caplets with strikes $\kappa_\ell(T_j)$ all maturing at T_j; work out their present values $\mathrm{Cpl}(0,\kappa_\ell(T_j),T_j)$.

Step-2: In the caplet formula (3.3.5), vary the shift $a(T_j)$ and zeta $\zeta(T_j)$ to *bestfit* the $\mathrm{Cpl}(0,\kappa_\ell(T_j),T_j)$, by minimizing

$$\sum_\ell \omega_\ell \left\| \begin{array}{c} \mathrm{Cpl}(0,\kappa_\ell(T_j),T_j) \\ -B(t,T_{j+1})\,\mathbf{B}\left(\begin{bmatrix} L(t,T_j)-\mu_j \\ +a(T_j)\end{bmatrix}, \begin{bmatrix}\kappa_\ell(T_j)-\mu_j \\ +a(T_j)\end{bmatrix}, \zeta(T_j)\right)\end{array}\right\|,$$

choosing the weights ω_ℓ to get a good fit at-the-money.

Step-3: Repeating Step-2 for T_j $(j = 1,2,..)$ produces the term structure of shift $a(T_j)$ and also the term structure of zeta $\zeta(T_j)$ (from which the volatilities $\xi(t,T)$ of the shifted forwards $H(t,T)$ must now be found).

To fit a *swaption skew* go through the same steps as for caplets using (3.4.16) to get the swaption style shift and zeta

$$\alpha_j = \sum_{\ell=j}^{N-1} u_\ell\left[a(T_\ell)-\mu_\ell\right], \qquad \zeta_j = \zeta(T_j,T_N) = \sqrt{\int_0^{T_j}\left|\sum_{\ell=j}^{\ell N-1} A_\ell\xi(t,T_\ell)\right|^2 dt}$$

for a set of swaptions that differ only in strike, that is, have the same exercise time T_j and same maturity T_N.

Varying j from 1 to $N-1$ produces the α_j $(j = 1,..,N-1)$ for a set of coterminal swaptions maturing at T_N, from which the termstructure of shift $a(T_\ell)$ $(\ell = 1,..,N-1)$ can be easily bootstrapped out. Similarly, fixing j and varying N produces a term structure of shift for a set of swaptions with the same exercise, while jointly varying j and N so the difference $(N-j)$ is fixed produces the term structure of shift for a set of swaptions with the same tenor.

Observe that the skew can only be fitted to a selection of data such as the caplets, or a subset of swaptions like a particular as a set of coterminal swaptions. At that point the shift $a(T)$ is fully determined and cannot be adjusted to fit other data. So to fit larger sets of volatility data, clearly some sort of averaging of shifts over an appropriate set of caplets and swaptions will be required.

7.2 Maturity only fit

A simple calibration to either caps or a diagonal of swaptions and also correlation, combines $c(T-t)$ and a maturity only dependent function like

$$\xi(t,T) = \phi(T)\, c(T-t) . \tag{7.1}$$

The simplicity of this calibration makes it effective and robust, and therefore suitable as a *default calibration* for more sophisticated routines that may fail for some reason.

Implied caplet volatilities can be directly inserted into $\phi(\cdot)$, because

$$\beta^2(T)\ T = \int_0^T |\xi(t,T)|^2\, dt = \int_0^T |\phi(T)\, c(T-t)|^2\, dt = \phi^2(T)\ T.$$

Fitting the implied volatilities $\beta(T_j, T_N)$ of $N-1$ coterminal swaptions exercising at T_j $(j = 1, .., N-1)$ and maturing at T_N requires

$$\beta^2(T_j, T_N)\ T_j = \int_0^{T_j} \left| \sum_{\ell=j}^{N-1} A_\ell^{(j,N)} \xi(t, T_\ell) \right|^2 dt,$$

$$= \sum_{\ell 1=j}^{N-1} \sum_{\ell 2=j}^{N-1} A_{\ell 1}^{(j,N)}\ A_{\ell 2}^{(j,N)} \phi_{\ell 1}\phi_{\ell 2} \int_0^{T_j} c^*(T_{\ell 1}-t)\, c(T_{\ell 2}-t)\, dt,$$

where the $\phi_\ell = \phi(T_\ell)$ can be bootstrapped as follows:

Step-1: For $j = 1, .., N-1$ and $\ell 1, \ell 2 = j, .., N-1$, precompute

$$\mathbf{X}_{\ell 1,\ell 2,j} = \int_0^{T_j} c^*(T_{\ell 1}-t)\, c(T_{\ell 2}-t)\, dt \quad \Rightarrow \quad \mathbf{X}_{\ell,\ell,j} = T_j \quad \textit{and express}$$

$$\beta^2(T_j, T_N)\ T_j = \sum_{\ell 1=j}^{N-1} \sum_{\ell 2=j}^{N-1} A_{\ell 1}^{(j,N)}\ A_{\ell 2}^{(j,N)} \phi_{\ell 1}\phi_{\ell 2} \mathbf{X}_{\ell 1,\ell 2,j} \qquad j = 1, .., N-1.$$

Step-2: Start the routine at $j = N-1$ setting $\phi_{N-1} = \beta(T_{N-1}, T_N)$.

Step-3: Knowing $\phi_{N-1}, .., \phi_{j+1}$, compute ϕ_j as the largest root $\mathbf{\Gamma}$ of

$$\left[A_j^{(j,N)}\right]^2 T_j\ \mathbf{\Gamma}^2 + 2\,\mathbf{\Gamma} \sum_{\ell 1=j+1}^{N-1} A_{\ell 1}^{(j,N)}\ A_j^{(j,N)} \phi_{\ell 1} \mathbf{X}_{\ell 1,j,j}$$

$$+ \left\{ \sum_{\ell 1=j+1}^{N-1} \sum_{\ell 2=j+1}^{N-1} A_{\ell 1}^{(j,N)}\ A_{\ell 2}^{(j,N)} \phi_{\ell 1}\phi_{\ell 2} \mathbf{X}_{\ell 1,\ell 2,j} - \beta^2(T_j, T_N)\ T_j \right\} = 0.$$

The calibration fails when roots are imaginary, for example, when implied volatilities fall steeply, but with normal market data that is unlikely.

REMARK 7.1 A frequently heard objection to this calibration is that the volatility function (7.1) is decidedly inhomogeneous. But by recasting $\phi(T) = \alpha + \phi_0(T)$ and then choosing the constant α to minimize

$$\sum_{j=1}^{N-1} \phi_0(T_k)^2 = \sum_{j=1}^{N-1} (\phi(T_k) - \alpha)^2 \quad \Rightarrow \quad \alpha = \frac{1}{N-1} \sum_{j=1}^{N-1} \phi(T_k),$$

it becomes clear that the calibration can, and usually does, have a substantial *homogeneous spine*

$$\xi(t,T) = \alpha\, c(T-t).$$

Note that here α is effectively the average of the $\phi(T_k)$, and that $\alpha\, c(T-t)$ is a sensible extension of the calibration to maturities beyond those for which data might be available. □

7.3 Homogeneous spines

Volatility functions that are entirely homogeneous

$$\xi(t,T) = \psi(T-t)\; c(T-t),$$

are attractive because of their stationary properties, but exact fits with them are difficult. We describe several that have been used in the past to bestfit caps and which might be used as *homogeneous spines* $\psi(T-t)$ in the more general volatility function

$$\xi(t,T) = \chi(t)\; \phi(T)\; \psi(T-t)\; c(T-t),$$

which can then be tweaked using the method of Section-7.5 to fit a selection of caplets and coterminal swaptions.

REMARK 7.2 A word of warning is that when bestfitted to caplets only, most of these homogeneous functions, other than perhaps the piecewise linear one, tend to produce functions with high values for small $x = T - t$ and low values for larger $x = T - t$, and they are therefore relatively useless for imparting information about swaption values relative to cap values. □

7.3.1 Piecewise linear

This is the sort of homogeneous volatility function used in the excellent Polypaths [97] mortgage backed securities (MBS) software, where it is bestfitted to a selection of liquid caps and swaptions. Select M nodes $T_1, T_2, ..., T_M$

(around six or seven is about right, spread out at say $1, 2, 5, 7, 10$ and 15 years), linearly interpolate between them and finish with some predetermined exponential decay λ, so that the homogeneous spine takes the form

$$\psi(T-t) = \psi(x) = \begin{bmatrix} a_0 \mathbb{I}[x=0] \\ +\sum_{i=1}^{M} \frac{\{(T_i - x) a_{i-1} + (x - T_{i-1}) a_i\}}{(T_i - T_{i-1})} \mathbb{I}[x \in (T_{i-1}, T_i]] \\ + \{a_{M+1} + (a_M - a_{M+1}) \exp(-\lambda [x - T_M])\} \mathbb{I}[x > T_M], \end{bmatrix}$$

which is determined by the $M+2$ constants $a_0, a_1, ..., a_M, a_{M+1}$. It is straightforward but messy, to get quadratic expressions in the $a_0, a_1, ..., a_M, a_{M+1}$ for caplet and swaption volatilities, which can then be bestfitted to a selection of liquid caps and swaptions.

7.3.2 Rebonato's function

Rebonato introduced the following homogeneous spine

$$\psi(T-t) = \psi(x) = (a + bx) \exp(-\lambda x) + c,$$

$$\Rightarrow \quad \beta^2(T)\ T = \int_0^T [(a+bx)\exp(-\lambda x) + c]^2\, dx$$

chosen to produce a hump (a defining feature of the forward volatility) at some appropriate point

$$x = \frac{1}{\lambda} - \frac{a}{b}.$$

7.3.3 Bi-exponential function

Here the homogeneous spine is

$$\begin{aligned} \psi(T-t) &= \psi(x) = a\exp(-\lambda x) + b\exp(-\mu x) + c, \\ \Rightarrow \quad & \beta^2(T)\ T = \int_0^T [a\exp(-\lambda x) + b\exp(-\mu x) + c]^2\, dx \\ &= a^2 F(2\lambda, T) + b^2 F(2\mu, T) + c^2 F(0, T) \\ &+ 2abF(\lambda+\mu, T) + 2acF(\lambda, T) + 2bcF(\mu, T) \quad \textit{where} \end{aligned}$$

$$F(\lambda, T) = \int_0^T \exp(-\lambda x)\, dx = \begin{cases} \frac{1}{\lambda}[1 - \exp(-\lambda T)] & \textit{if} \quad \lambda > 0, \\ T \quad \textit{if} \quad \lambda = 0. \end{cases}$$

7.3.4 Sum of exponentials

Given a selection of M (around 5 to 7) caplets maturing at times $T_1, T_2, ..., T_M$ (spread out, like $1, 2, 5, 7, 10, 15$) with implied volatilities $\beta_1, \beta_2, ..., \beta_M$ the following algorithm will exactly fit a homogeneous spine of form

$$\psi(T-t)=\psi(x)=\sum_{j=1}^{M}a_j\exp(-\lambda_j x)\qquad \lambda_j=j\,\frac{\lambda}{M}\qquad \lambda\cong 3.$$

Assuming $\beta_{i+1}^2T_{i+1}\geq\beta_i^2T_i$, evaluate the *caplet interval values* V_i analytically and also apply the mean value theorem to get for $i=0,1,...,M-1$

$$V_i=\sqrt{\beta_{i+1}^2T_{i+1}-\beta_i^2T_i}=\sqrt{\int_{T_i}^{T_{i+1}}\psi^2(x)\,dx}$$
$$=\sqrt{\sum_{j=1}^{M}\sum_{k=1}^{M}\frac{a_ja_k}{(\lambda_j+\lambda_k)}\left[\mathbf{e}^{-(\lambda_j+\lambda_k)T_i}-\mathbf{e}^{-(\lambda_j+\lambda_k)T_{i+1}}\right]},\tag{7.2}$$

$$\exists\ c_i\in(T_i,T_{i+1})\quad\text{such that}\quad V_i=\sum_{j=1}^{M}a_j\exp(-\lambda_jc_i).\tag{7.3}$$

Step-1: Set $c_i=\frac{1}{2}(T_i+T_{i+1})$ and solve (7.3) for the corresponding a_j.

Step-2: Substitute the a_j $(j=1,...,M)$ into (7.2) to obtain caplet interval values $\widetilde{V}_i=V_i+\Delta V_i$ $(i=0,1,...,M-1)$ corresponding to the a_j.

Step-3: Adjust the current mean value variables in equations (7.3) $c_i\to c_i+\Delta c_i$ so that with the *current* a_j

$$\sum_{j=1}^{M}a_j\exp(-\lambda_j[c_i+\Delta c_i])=V_i-\Delta V_i\quad\Rightarrow\quad\Delta c_i\cong\frac{\Delta V_i}{\sum_{j=1}^{M}a_j\lambda_j\exp(-\lambda_jc_i)}.$$

Step-4: In the equations (7.3) replace c_i by $c_i+\Delta c_i$ as computed in Step-3, solve them for a new set of a_j, and return to Step-2.

The algorithm works quite well in practice, but the author has not established a sufficient condition for convergence.

7.4 Separable one-factor fit

Volatility functions which are one-factor and *separable*

$$\xi(t,T)=\phi(T)\,\chi(t),$$

are not homogeneous and carry no correlation information, yet are flexible enough to jointly parametrize caps and a diagonal of coterminal swaptions.

The implied volatilities of a caplet and swaption exercising at T_j, are

$$\beta^2(T_j)\ T_j = \phi^2(T_j)\int_0^{T_j}\chi(t)^2\,dt \qquad \textit{and} \tag{7.4}$$

$$\beta^2(T_j,T_N)\ T_j = \left(\sum_{\ell=j}^{N-1} A_\ell^{(j,N)}\phi(T_\ell)\right)^2\int_0^{T_j}\chi(t)^2\,dt$$

respectively. The task is to fit $N-1$ caplets exercising at $T_1,..T_{N-1}$, and $N-1$ coterminal swaptions exercising at $T_1,..T_{N-1}$ and settling at T_N, altogether $2N-3$ instruments (the last caplet and swaption coincide).

Put $\phi(T_j)=\phi_j$ for $j=1,..,N-1$, assume $\chi(\cdot)$ is piecewise constant on each interval like

$$\chi(t)=\chi_k \quad \textit{for} \quad t\in(T_{k-1},T_k] \quad k=1,..,N \qquad \textit{and set} \tag{7.5}$$

$$\mathbf{X}_j=\mathbf{X}(T_j)=\int_0^{T_j}\chi^2(t)\,dt=\sum_{k=1}^{j}\Delta_k\ \chi_k^2, \qquad j=1,..,N-1.$$

With this notation, equations (7.4) become

$$\beta^2(T_j)\,T_j=\phi_j^2\mathbf{X}_j,\quad \beta^2(T_j,T_N)\,T_j=\left(\sum_{\ell=j}^{N-1}A_\ell^{(j,N)}\phi_\ell\right)^2\mathbf{X}_j, \tag{7.6}$$

and a successful solution to the calibration problem would be to identify the $2N-2$ positive parameters

$$\mathbf{X}_1<\mathbf{X}_2<...,<\mathbf{X}_{N-1};\ \phi_1,\ \phi_2,..,\ \phi_{N-1}$$

that return the required caplet and swaption volatilities. Work backwards from the final exercise.

Step-1: Because the expressions (7.6) involve products of $\chi(\cdot)$ and $\phi(\cdot)$, instability arising from *cross-leakage* between the two functions is possible. To prevent this we fit the last caplet/swaption by choosing

$$\mathbf{X}_{N-1}=T_{N-1},\quad \phi_{N-1}=\beta(T_{N-1})=\beta(T_{N-1},T_N).$$

Step-2: Continuing backwards step-by-step, suppose $\mathbf{X}_{N-1},..,\mathbf{X}_{j+1}$ and $\phi_{N-1},..,\phi_{j+1}$ satisfy (7.6) at $T_{N-1},..,T_{j+1}$. To satisfy (7.6) at T_j, divide the two equations to eliminate $\mathbf{X}_j$ and make ϕ_j the largest root Γ of

$$\mathbf{\Gamma}^2\ \beta^2(T_j,T_N)=\beta^2(T_j)\left(\mathbf{\Gamma}\ A_j^{(j,N)}+\sum_{\ell=j+1}^{N-1}A_\ell^{(j,N)}\phi_\ell\right)^2.$$

As long as $\beta(T_j, T_N) > A_j^{(j,N)} \beta(T_j)$ (an inequality difficult to break with market data) there is one positive root, so

$$\phi_j = \frac{\beta(T_j) \sum_{\ell=j+1}^{N-1} A_\ell^{(j,N)} \phi_\ell}{\beta(T_j, T_N) - A_j^{(j,N)} \beta(T_j)} \qquad \Rightarrow \qquad \mathbf{X}_j = \frac{\beta^2(T_j)}{\phi_j^2} T_j.$$

Step-3: Assuming that at each backward step $\beta(T_j, T_n) > A_j^{(j,N)} \beta(T_j)$, we will have consecutively computed ϕ_j and $\mathbf{X}_j$ for $k = N-1, .., 1$. Then, so long as the $\mathbf{X}_j$ are monotonic increasing, the calibration is completed by retrieving positive

$$\chi_1 = \sqrt{\frac{\mathbf{X}_1}{\Delta_1}}, \qquad \chi_j = \sqrt{\frac{\mathbf{X}_j - \mathbf{X}_{j-1}}{\Delta_j}}, \quad j = 2, .., N-1.$$

Producing an $\mathbf{X}(T)$ that is not monotonic increasing (so $\chi(T)$ becomes imaginary for some T) is how the model reacts to a potentially arbitragable situation (like when swaption volatilities are unreasonably greater than caplet volatilities). Otherwise the algorithm is extremely robust; for example, it can parametrize caplet volatilities starting at 40% and converging at 10-years maturity to a diagonal of swaption volatilities each of the order of 20%.

REMARK 7.3 Even without correlation, this calibration got satisfactory prices for callable range rate accruals, in which the range rate part requires caplet volatility, and the callability part requires coterminal swaption volatility. First the skew was fitted to return caplet volatilities at the strikes banding the accrual, and then this algorithm fitted caplet volatilities at one of those strikes and also coterminal swaption volatilities at-the-money. □

7.5 Separable multi-factor fit

The 1-factor separable algorithm can be articulated to a *separable multi-factor* setting in which correlation, caplets and swaptions have already been *best-fitted*, either with a homogeneous spine as in Section-7.3 or with a method like Pedersen's of Section-7.6, but an exact fit to a selection of instruments is now required.

Assume therefore that we already have a vector volatility function $\psi(t, T)$ that incorporates correlation and fits a selection of instruments, and concentrate on finding the scalar functions $\phi(T)$ and $\chi(t)$ in

$$\xi(t, T) = \phi(T) \chi(t) \psi(t, T) = \phi(T) \chi(t) \begin{pmatrix} \psi^{(1)}(t, T) \\ \psi^{(2)}(t, T) \\ \vdots \end{pmatrix}$$

to exactly fit a strip of caplets and diagonal of coterminal swaptions as in Section-7.4. The idea in the following algorithm is to vary *dummy* swaption volatilities in the 1-factor routine of Section-7.4, so that when the *true* ones in this multi-factor routine are recomputed they become ever closer to the required ones.

So similarly to Section-7.4, our task is to find $\chi(t)$ and $\phi(T)$ to fit $N-1$ caplet and $N-1$ swaption implied volatilities at exercise timesT_j (for $j = 1, 2, .., N-1$) like

$$\beta^2(T_j)T_j = \phi^2(T_j)\int_0^{T_j}\chi^2(t)|\psi(t,T_j)|^2\,dt, \tag{7.7}$$

$$\beta^2(T_j,T_N)T_j = \sum_{\ell1=j}^{N-1}\sum_{\ell2=j}^{N-1}\left\{\begin{array}{c} A_{\ell1}^{(j,N)}A_{\ell2}^{(j,N)}\phi(T_{\ell1})\phi(T_{\ell2}) \\ \times\int_0^{T_j}\chi^2(t)\psi^*(t,T_{\ell1})\psi(t,T_{\ell2})\,dt \end{array}\right\}.$$

Because the expression frequently recurs and can be precomputed (perhaps analytically and exactly) from the given $\psi(t,T)$, introduce

$$\Psi[\ell1,\ell2,k] = \frac{1}{\Delta T_k}\int_{T_{k-1}}^{T_k}\psi^*(t,T_{\ell1})\psi(t,T_{\ell2})\,dt$$

so that, if $\chi(t) = \chi_k$ for $t\in(T_{k-1},T_k]$, that is, $\chi(\cdot)$ is piecewise constant as in (7.5), then for $j = 1, .., N-1$ and $\ell1, \ell2 \geq j$ we can define

$$\mathbf{X}_{\ell1,\ell2,j} = \int_0^{T_j}\chi^2(t)\psi^*(t,T_{\ell1})\psi(t,T_{\ell2})\,dt = \sum_{k=1}^{j}\chi_k^2\Psi[\ell1,\ell2,k]\Delta_k$$

$$\mathbf{X}_j = \mathbf{X}_{j,j,j} = \int_0^{T_j}\chi^2(t)|\psi(t,T_j)|^2\,dt = \sum_{k=1}^{j}\chi_k^2\Psi[j,j,k]\Delta_k$$

in terms of χ_k and $\Psi[\cdot]$. Also, knowing the $\mathbf{X}_j$, the corresponding $\chi_j \quad j = 1, .., N-1$ can be consecutively recovered, so long as they are real, from

$$\mathbf{X}_1 = \chi_1^2\ \Psi[1,1,1]\Delta T_1 \quad \textit{and for} \quad j = 2, .., N-1 \quad \textit{from} \tag{7.8}$$

$$\mathbf{X}_j = \chi_j^2\ \Psi[j,j,j]\Delta T_j + \sum_{k=1}^{j-1}\chi_k^2\Psi[j,j,k]\Delta T_k.$$

Setting $\phi(T_j) = \phi_j$, from (7.7) our task in the multi-dimensional separable case is (as in Section-7.4) to identify $\phi_1, .., \phi_{N-1}$ and $\mathbf{X}_1, .., \mathbf{X}_{N-1}$ that return the required caplet and swaption volatilities for $j = 1, .., N-1$

$$\beta^2(T_j)T_j = \phi_j^2\mathbf{X}_j, \tag{7.9}$$

$$\beta^2(T_j,T_N)T_j = \sum_{\ell1=j}^{N-1}\sum_{\ell2=j}^{N-1}A_{\ell1}^{(j,N)}A_{\ell2}^{(j,N)}\phi_{\ell1}\phi T_{\ell2}\mathbf{X}_{\ell1,\ell2,j}.$$

The following iterative routine to find $\phi_1, .., \phi_{N-1}$ and $\mathbf{X}_1, .., \mathbf{X}_{N-1}$ cycles the 1-factor routine of Section-7.4, and appears quite robust (converging to 8 or 9 decimal places of the target implied volatilities in about 20 loops in the authors R&D program).

Step-1: To avoid *cross leakage* between $\chi(\cdot)$ and $\phi(\cdot)$, lock ϕ_{N-1} and $\mathbf{X}_{N-1}$ to the $(N-1)^{th}$ caplet/swaption volatility via

$$\mathbf{X}_{N-1} = \int_0^{T_{N-1}} |\psi(t, T_{N-1})|^2 \, dt \qquad \beta^2(T_{N-1}) T_{N-1} = \phi^2(T_{N-1}) \mathbf{X}_{N-1}$$

Step-2: Starting with the already specified ϕ_{N-1} and $\mathbf{X}_{N-1}$ use the 1-factor routine of Section-7.4 to obtain the $\phi_1, .., \phi_{N-2}$ and $\mathbf{X}_1, .., \mathbf{X}_{N-2}$ solving the *subsidiary problem*

$$\beta^2(T_j) T_j = \phi_j^2 \mathbf{X}_j,$$

$$\widetilde{\beta}^2(T_j, T_N) T_j = \sum_{\ell 1=j}^{N-1} \sum_{\ell 2=j}^{N-1} A_{\ell 1}^{(j,N)} A_{\ell 2}^{(j,N)} \phi_{\ell 1} \phi_{\ell 2} \mathbf{X}_j,$$

in which the *target* swaption volatilities $\widetilde{\beta}(T_j, T_N)$ are the *true* ones $\widetilde{\beta}(T_j, T_N)$ $= \beta(T_j, T_N)$ during the first passage, but are changed for subsequent passages.

Step-3: Using (7.8) recover from $\mathbf{X}_1, .., \mathbf{X}_{N-1}$ the corresponding $\chi_1, ., \chi_{N-1}$.

Step-4: Insert the $\phi_1, .., \phi_{N-1}$ and $\chi_1, .., \chi_{N-1}$ computed in Steps-2,3 into the *true expression* (7.9), and obtain the swaption volatility error ε_j (note $\varepsilon_{N-1} = 0$ always because of Step-1) satisfying

$$\left[\beta^2(T_j, T_N) - \varepsilon_j\right] T_j = \sum_{\ell 1=j}^{N-1} \sum_{\ell 2=j}^{N-1} A_{\ell 1}^{(j,N)} A_{\ell 2}^{(j,N)} \phi_{\ell 1} \phi T_{\ell 2} \mathbf{X}_{\ell 1, \ell 2, j}.$$

Step-5: Return to Step-2, replacing the target swaption volatilities $\widetilde{\beta}^2(T_j, T_N)$ with $\widetilde{\beta}^2(T_j, T_N) + \varepsilon_j$ (when the routine works, the errors $\varepsilon_j \to 0$ rapidly, forcing the target swaptions into the real ones).

REMARK 7.4 In the totally inhomogeneous case when $\psi(t, T) = 1$, Step-1 gives $\phi(T)$ the flavour of volatility and $\chi(t)$ the flavour of perturbation away from 1. In contrast, when $\psi(\cdot)$ is a bestfit, Step-1 lends both $\phi(T)$ and $\chi(t)$ the flavour of perturbations away from 1. Also, while the algorithm works quite well in practice, the author has not established sufficient conditions for convergence. □

7.5.1 Alternatively

An alternative approach is an iterative routine, on each loop first fitting the swaptions by varying the $\chi(t)$ and then fitting the caplets by varying the $\phi(T)$. Specifically:

Step-1: To avoid *cross leakage* between $\chi(\cdot)$ and $\phi(\cdot)$, lock $\phi(T_{N-1}) = \phi_{N-1}$ and $\mathbf{X}_{N-1}$ to the $(N-1)^{th}$ caplet/swaption volatility via

$$\mathbf{X}_{N-1} = \int_0^{T_{N-1}} |\psi(t, T_{N-1})|^2 \, dt \qquad \beta^2(T_{N-1}) T_{N-1} = \phi^2(T_{N-1}) \mathbf{X}_{N-1},$$

and then set $\phi(T_j) = \phi_j = 1$ for $j = 1, 2, ..., N-2$.

Step-2: With a given ϕ_j, use the swaption data

$$\beta^2(T_j, T_N) T_j = \sum_{\ell 1=j}^{N-1} \sum_{\ell 2=j}^{N-1} A_{\ell 1}^{(j,N)} A_{\ell 2}^{(j,N)} \phi_{\ell 1} \phi T_{\ell 2} \mathbf{X}_{\ell 1, \ell 2, j},$$

$$\mathbf{X}_{\ell 1, \ell 2, j} = \sum_{k=1}^{j} \chi_k^2 \Psi[\ell 1, \ell 2, k] \Delta_k$$

for $j = 1, 2, ..$to successively back out the χ_j^2, setting $\chi_j^2 = 0$ if the required number should be negative.

Step-3: With the χ_j^2 from the previous step, use the caplet data

$$\beta^2(T_j) T_j = \phi_j^2 \mathbf{X}_j, \qquad \mathbf{X}_j = \mathbf{X}_{j,j,j} = \sum_{k=1}^{j} \chi_k^2 \Psi[j, j, k] \Delta_k,$$

to successively back out the ϕ_j for $j = 1, 2, ..., N-2$. Then return to Step-2 and iterate.

REMARK 7.5 Rather than bootstrapping to a full spectrum of caplets and coterminal swaptions, this method can clearly be adapted to fitting only liquid ones by using an optimizer at each step to hit the liquid instruments while smoothing the ϕ_j and χ_j^2 with respect to j. □

7.6 Pedersen's method

The *generic calibration* described in this section is based on Pedersen's paper [85], which the author recommends consulting for its numerous practical implementation details.

Given an optimizer like the NAG [81] or IMSL [64] ones (or Powell's method from Press's Numerical Recipes [98] as suggested by Pedersen), his algorithm bestfits a selection of liquid caplets and swaptions, incorporates historical correlation and produces a smooth r-factor *pseudo-homogeneous volatility function* of the form

$$\xi(t, x) = \xi(t, T)|_{T=t+x} \, .$$

We assume the $(N-1)\times(N-1)$ matrix from which the target caplet and swaptions ($\text{target}_i \quad i=1,2,..$) are drawn has shifted BGM implied volatilities entered by quarterly exercise in the rows, and quarterly increasing tenors in the columns, after the fashion of Example-7.1.

The $(N-1)\times(N-1)$ matrix X that will be varied by the optimizer in this algorithm, comprises $(N-1)^2$ parameters,

$$X = (X_{k,\ell}) \qquad k,\ell = 1,..,N-1$$

with each parameter $X_{k,\ell}$ approximately the magnitude $|\xi(t,x)|$ of the vector volatility function $\xi(t,x)$ on certain intervals. Specifically we suppose $\xi(t,x)$ is piecewise constant with

$$|\xi(t,x)| \cong X_{k,\ell} \qquad t\in(T_{k-1},T_k], \quad x\in(x_{\ell-1},x_\ell] \quad k,\ell=1,..,N-1.$$

To tolerate this inexactness, bear in mind that in this routine X is simply a vehicle to producing a volatility function $\xi(t,x)$ by the slightly roundabout process clearly specified below, and is not particularly significant in itself.

The *objective function* for the optimizer involves several components and is of form

$$\text{obj}(X) = w\ \text{bestfit}(X) + w_1\ \text{smooth}^{(1)}(X) + w_2\ \text{smooth}^{(2)}(X), \qquad (7.10)$$

where the weights w and w_1, w_2 are chosen to get a balance between *bestfit* to the target caplets and swaptions, and *smoothness* of X in both row and column directions. Both the fit and smoothing functions

$$\text{bestfit}(X) = \sum_i \left|\frac{\text{swpn}_i(X)}{\text{target}_i} - 1\right|^2,$$

$$\text{smooth}^{(1)}(X) = \sum_{k=1}^{N-1}\sum_{\ell=2}^{N-1}\left|\frac{X_{k,\ell}}{X_{k,\ell-1}} - 1\right|^2,$$

$$\text{smooth}^{(2)}(X) = \sum_{k=2}^{N-1}\sum_{\ell=1}^{N-1}\left|\frac{X_{k,\ell}}{X_{k-1,\ell}} - 1\right|^2,$$

are designed to be independent of *scale.*

From X and the historical correlation function $c(\cdot)$ (see Chapter-6), the swaption prices $\text{swpn}_i(X)$ are computed by first constructing a volatility function $\xi(\cdot)$ and then pricing in the standard fashion. For simplicity, fixed and floating side nodes are assumed to coincide with $T_j = \delta j$ for $j=0,1,..,N$.

For each k corresponding to *the* k^{th} *slice of calendar time* $t\in(T_{k-1},T_k]$, given X the $(N-1)\times(N-1)$ matrix (with $\ell1,\ell2 = 1,..,N-1$)

$$C_k = \left(X_{k,\ell1}\ c(x_{\ell1})^T c(x_{\ell2})\ X_{k,\ell2}\right)$$

is a covariance, because the correlation of the relative shifted forwards $H(t, x_{\ell 1})$ and $H(t, x_{\ell 2})$ is $c^T(x_{\ell 1})\, c(x_{\ell 2})$. So if we eigenvalue decompose C_k, let $\Gamma^{(1)}$, ..,$\Gamma^{(r)}$ be the first r eigenvectors ordered by eigenvalue size, multiplied by the square root of the corresponding eigenvalues, and introduce the $(N-1) \times r$ matrix

$$\Gamma = \left(\Gamma^{(1)}, ..., \Gamma^{(r)}\right) \qquad \textit{then} \qquad \Gamma^T \Gamma \cong C_k.$$

The last equation is approximate because we have discarded some eigenvectors, and that is the reason $X_{k,\ell}$ is only roughly the magnitude of $|\xi(t,x)|$. Now construct a volatility function $\xi(\cdot)$ for the kth slice of calendar time $t \in (T_{k-1}, T_k]$ corresponding to the current value of X by setting the vector volatility function

$$\xi(t,x) = \xi_{k.\ell} = \left(\Gamma_\ell^{(1)}, ..., \Gamma_\ell^{(r)}\right) \quad \textit{for} \quad x \in (x_{\ell-1}, x_\ell].$$

The full vector volatility function $\xi(t,x) = \xi_{k.\ell}$ covering all calendar times t simply requires this step to be repeated for each slice of calendar time $t \in (T_{k-1}, T_k]$ for $k = 1, 2, ..$

Note that if $t \in (T_{k-1}, T_k]$ the relative maturity x corresponding to T_j is bounded, that is

$$t \in (T_{k-1}, T_k] \quad \textit{and} \quad x + t = T_j \quad \Rightarrow \quad x \in [T_j - T_k,\ T_j - T_{k-1}).$$

That means for each quarterly block of calendar time, the volatility $\xi(t, T_j)$ at absolute maturity T_j is also constant because

$$\xi(t, T_j)|_{T_j = t+x} = \xi(t, x_{j-k+1}) = \xi_{k,\ j-k+1} \quad \textit{for} \quad t \in (T_{k-1}, T_k].$$

Caplet and swaption implied volatilities are then easily computed from (4.16), because they are a linear combination of integrals of type

$$\begin{aligned}\int_0^T \xi^*(t, T_{j1})\, \xi(t, T_{j2})\, dt &= \sum_k \int_{T_{k-1}}^{T_k} \xi^*(t, T_{j1})\, \xi(t, T_{j2})\, dt \\ &= \delta \sum_k \xi^*_{k,\ j1-k+1}\, \xi_{k,\ j2-k+1}.\end{aligned}$$

REMARK 7.6 In general we will not have

$$\Gamma = \left(\Gamma_\ell^{(1)}, ..., \Gamma_\ell^{(r)}\right) = \left(X_\ell\ c(x_\ell)\right),$$

that is, the eigenvector decomposition step is not redundant because the components of $(c(x_\ell))$ are usually different from the eigenvectors of C_k. Moreover, in practice the eigenvector decomposition applied to each slice of calendar time, appears to induce considerable variability in the vector components of

$\xi(\cdot)$. An implementation of the algorithm viewed by the author had the X matrix represented by cells on an Excel spreadsheet with the cell colour changing with size of the corresponding component; the resulting filmshow as X converged to the desired degree of bestfit and smoothness was both entertaining and educational. The author feels Pedersen's algorithm works well precisely because it can shuffle volatility between rows and then between factors within rows in an efficient and coherent way. □

The overall modus-operandi for Pedersen's algorithm is thus:

Step-1 Start X either with *yesterday's* X, or with magnitudes of the historical volatility function $\xi(x)$ by setting $X_{k,\ell} = |\xi(x_\ell)|\ \forall k$.

Step-2 From X compute the implied volatilities of the target caplets and swaptions using the process described above.

Step-3 Insert those values along with X and the target implied volatilities $\text{target}_i \quad i = 1, 2, ..$ into the objective function (7.10) and feed into the optimizer to generate a new trial X.

Step-4 Return to Step-2 and iterate until the desired fit and degree of smoothness are obtained. Then extract the final $\xi(t, x)$ as the desired *bestfit* volatility function.

7.7 Cascade fit

This *generic calibration* attempts to fit all caps, all swaptions and correlation with a piecewise constant volatility function of type

$$\xi(t,T) = \gamma(t,T)\, c(T-t), \quad \gamma(t,T_\ell) = \gamma_{k,\ell} \quad t \in (T_{k-1}, T_k] \quad k = 1, 2, .., n$$

for which all coverages are uniform $\delta_j = \overline{\delta}_i = \delta$ vis-a-vis the nodes $T_0 = 0, T_1, ..., T_n$, and the parameters $\gamma_{k,\ell}$ for $\ell \geq k \geq 1$ define the calibration.

REMARK 7.7 The algorithm will require solving many quadratic equations which in practice frequently produce imaginary roots. So it frequently fails at a significant number of points where the corresponding $\gamma_{j,\ell}$ must be made small or zero. The result is an unattractive volatility function with many peaks and troughs, a problem which Brigo and Mercurio [32] showed could be partially tackled by smoothing the cap/swaption volatility surface. □

Assume for $1 \leq j < N \leq n$ we are given the $\frac{1}{2}n(n-1)$ caplet and swaption volatilities

$$\beta^2(T_j,T_N)\ T_j = \int_0^{T_j}\left|\sum_{\ell=j}^{N-1} A_\ell^{(j,N)}\xi(t,T_\ell)\right|^2 dt$$

$$= \sum_{\ell 1=j}^{N-1}\sum_{\ell 2=j}^{N-1} A_{\ell 1}^{(j,N)}\ A_{\ell 2}^{(j,N)} \int_0^{T_j}\gamma(t,T_{\ell 1})\gamma(t,T_{\ell 2})\,c^*(T_{\ell 1}-t)\,c(T_{\ell 2}-t)\,dt,$$

where the weights $A_\ell^{(j,N)}$ $(j\le \ell < N)$ refer to the swaption maturing at T_j with final settlement at T_N, and the caplets occur at $j = N-1$ when $A_{N-1}^{(N-1,N)} = 1$. The integral term equals

$$\sum_{k=1}^{j}\gamma_{k,\ell 1}\ \gamma_{k,\ell 2}\int_{T_{k-1}}^{T_k} c^*(T_{\ell 1}-t)\,c(T_{\ell 2}-t)\,dt = \delta\sum_{k=1}^{j}\gamma_{k,\ell 1}\ \gamma_{k,\ell 2}\ \mathbf{X}_{\ell 1,\ell 2,k}$$

in which we can precompute the correlation terms

$$\mathbf{X}_{\ell 1,\ell 2,k} = \frac{1}{\delta}\int_0^{\delta} c^*(T_{\ell 1}+T_{k-1}-t)\,c(T_{\ell 2}+T_{k-1}-t)\,dt \quad\Rightarrow\quad \mathbf{X}_{\ell,\ell,k} = 1.$$

For $1\le j<N\le n$ the task therefore becomes finding $\gamma_{k,\ell}$ to satisfy

$$j\ \beta^2(T_j,T_N) = \sum_{k=1}^{j}\sum_{\ell 1=j}^{N-1}\sum_{\ell 2=j}^{N-1} A_{\ell 1}^{(j,N)}\ A_{\ell 2}^{(j,N)}\gamma_{k,\ell 1}\ \gamma_{k,\ell 2}\ \mathbf{X}_{\ell 1,\ell 2,k}. \tag{7.11}$$

Given j and $\gamma_{k,\ell}$ $(k<j,\ k\le \ell\le n-1)$, $\gamma_{j,\ell}$ $(j\le \ell\le N-2)$, the *cascade* routine satisfies (7.11) by making $\gamma_{j,N-1}$ the largest root Γ of

$$\left(A_{N-1}^{(j,N)}\right)^2\ \Gamma^2 + 2\,\Gamma\sum_{\ell 1=j}^{N-2} A_{\ell 1}^{(j,N)}\ A_{N-1}^{(j,N)}\gamma_{j,\ell 1}\ \mathbf{X}_{\ell 1,N-1,j} \tag{7.12}$$

$$+\left\{\begin{array}{c}\sum_{\ell 1=j}^{N-2}\sum_{\ell 2=j}^{N-2} A_{\ell 1}^{(j,N)}\ A_{\ell 2}^{(j,N)}\gamma_{j,\ell 1}\ \gamma_{j,\ell 2}\ \mathbf{X}_{\ell 1,\ell 2,j}\\ +\sum_{k=1}^{j-1}\sum_{\ell 1=j}^{N-1}\sum_{\ell 2=j}^{N-1}\left[\begin{array}{c}A_{\ell 1}^{(j,N)}\ A_{\ell 2}^{(j,N)}\gamma_{k,\ell 1}\ \gamma_{k,\ell 2}\ \mathbf{X}_{\ell 1,\ell 2,k}\\ -j\ \beta^2(T_j,T_N)\end{array}\right]\end{array}\right\} = 0.$$

For $j=1$, start off the cascade (7.12) with $\gamma_{1,1} = \beta(T_1,T_2)$ and consecutively determine $\gamma_{1,2}, \gamma_{1,3},\dots$ When $j>1$ set $N=j+1$ in (7.11) and make $\gamma_{j,j}$ the positive root Γ of

$$\Gamma^2 + \left\{\sum_{k=1}^{j-1}(\gamma_{k,j})^2 - j\ \beta^2(T_j,T_{j+1})\right\} = 0,$$

to set off the cascade (7.12) with $\gamma_{j,j}$ and consecutively determine $\gamma_{j,j+1}$, $\gamma_{j,j+2}\dots$

7.7.1 Extension

A useful extension of the cascade algorithm is to attempt an exact fit after first obtaining a bestfit. For example, suppose we have bestfitted correlation and the swaption matrix by the Pedersen method and produced a vector volatility function $\psi(t,T)$. We can now attempt an *exact fit* with the cascade method using a volatility function of type

$$\xi(t,T) = \psi(t,T) + \gamma(t,T)\,c(T-t),$$
$$\gamma(t,T_\ell) = \gamma_{k,\ell} \quad t \in (T_{k-1}, T_k] \quad k = 1,2,..,n$$

where the parameters $\gamma_{k,\ell}$ for $\ell \geq k \geq 1$ define the new calibration.

As above, for $1 \leq j < N \leq n$ and the given $\frac{1}{2}n(n-1)$ caplet and swaption volatilities, that means satisfying the equations

$$\beta^2(T_j, T_N)\ T_j = \int_0^{T_j} \left| \sum_{\ell=j}^{N-1} A_\ell^{(j,N)} \left[\psi(t,T_\ell) + \gamma(t,T_\ell)\,c(T_\ell - t)\right] \right|^2 dt,$$

which generates a system of quadratic equations like (7.12) that can be solved in a similar consecutive fashion.

7.8 Exact fit with semidefinite programming

Starting with the same kind of pseudo-homogenous *volatility function*

$$\xi(t,x) = \xi(t,T)|_{T=t+x} = \xi_{k.\ell} \quad \textit{when} \quad t \in (T_{k-1}, T_k] \quad x \in (x_{\ell-1}, x_\ell]$$

as in Pedersen's approach (see Section-7.6 above), and also varying correlation (which moves swaption but not caplet volatilities), it is possible to get an exact fit to the swaption volatility matrix under quite adverse market conditions (as during the collapse of Longterm Capital when short-term volatilities were very high) using *semidefinite programming* (SDP), see Appendix-A.4.

Using *homogenous by layer* type volatility functions like ours, swaption zetas (that is, implied volatility squared multiplied by time to maturity, see 4.16) turn out to be linear combinations of covariance matrices, leading to linear equality constraints in the SDP framework. Moreover, because $\xi(\cdot)$ is pseudo-homogenous, instantaneous correlation can be sensibly defined, allowing us to construct an objective function to bestfit historic covariance in a certain sense, and think of the resulting correlation structure as *implied correlation.*

Introduce $N-1$ semidefinite matrices $X^{(k)} \in \mathbb{S}_{N-k}$ for $k = 1,2,\ldots,N-1$ (so $X^{(1)} \in \mathbb{S}_{N-1}, \ldots, X^{(N-1)} \in \mathbb{S}_1$) defined by

$$X^{(k)} = \left(X^{(k)}_{\ell 1,\,\ell 2}\right) \qquad X^{(k)}_{\ell 1,\,\ell 2} = \xi^*_{k,\,\ell 1}\,\xi_{k,\,\ell 2}, \quad \ell 1, \ell 2 = 1,2,\ldots,N-k,$$

and with them form the $\frac{1}{2}N(N-1) \times \frac{1}{2}N(N-1)$ block semidefinite matrix

$$X = \operatorname{diag}\left(X^{(1)}, .., X^{(N-1)}\right)$$

Swaption implied volatilities are linear combinations of integrals like

$$\int_0^{T_m} \xi^*(t, T_{j1})\, \xi(t, T_{j2})\, dt = \sum_{k=1}^{m} \int_{T_{k-1}}^{T_k} \xi^*(t, T_{j1})\, \xi(t, T_{j2})\, dt$$
$$= \delta \sum_{k=1}^{m} \xi_{k,\ j1-k+1}\ \xi_{k,\ j2-k+1} = \delta \sum_{k=1}^{m} X^{(k)}_{j1-k+1,\ j2-k+1},$$

which in turn are linear combinations of the elements of X. Hence from (4.16), the implied volatility $\beta(T_m, T_n)$ of a swaption exercising at T_m and maturing at T_n can be expressed in the form of a standard SDP equality constraint

$$\beta^2(T_m, T_n)\, T_m = U^{(m,n)} \cdot X \tag{7.13}$$
$$= \sum_{j1=m}^{n-1} \sum_{j2=m}^{n-1} A^{(m,n)}_{j1} A^{(m,n)}_{j2} \delta \sum_{k=1}^{m} X^{(k)}_{j1-k+1,\ j2-k+1}.$$

Our aim now is to satisfy swaption implied volatility constraints like (7.13) using an SDP objective function that produces an X which embodies a satisfactory implied correlation structure. In addressing that task our *basic assumptions* (which seem to have become fairly accepted wisdom) are:

1. Ordering by eigenvalue size, the principle components $\left\{e^{(i)}\right\}$ of the historical covariance matrix are stable, but the level of volatility, that is, the corresponding eigenvalues $\{\lambda_i\}$, may change.

2. Some 3 to 5 factors and no more, are needed to adequately explain movement of the simple forward curves.

That suggests a model for covariance like

$$\sum_{i=1}^{5} \lambda_i e^{(i)}\, e^{(i)*},$$

with the positive eigenvalues λ_i made variable, but the eigenvectors $e^{(i)}$ held constant.

For each $i = 1, .., 5$, introduce the $N-1$ matrices $E^{(i,k)} \in \mathbb{S}_{N-k}$ for $k = 1, 2, \ldots, N-1$ defined by

$$E^{(i,1)} = e^{(i)}\, e^{(i)*}, \quad E^{(i,k)} = \left(e^{(i)}_j : j = 1, .., N-k\right)\left(e^{(i)}_j : j = 1, .., N-k\right)^*$$

that is, each $E^{(i,k)}$ is the previous $E^{(i,k-1)}$ less its last row and column. We make a linear combination of the $E^{(i,k)}$ the *target* for the for the covariance

$X^{(k)}$ in the kth layer. Specifically, for each $k = 1, .., N-1$ seek semidefinite $X^{(k)}$ and numbers $\lambda_i^{(k)} \geq 0$ $(i = 1, .., 5)$ to minimize

$$\left\| X^{(k)} - \sum_{i=1}^{5} \lambda_i^{(k)} E_i^{(k)} \right\|_2 .$$

That can be done by solving the *optimization problem*

find	$X^{(k)},\ \lambda_i^{(k)} \quad i = 1, .., 5;\ k = 1, .., N-1$
to minimize	$\sum_{k=1}^{N-1} \left\| X^{(k)} - \sum_{i=1}^{5} \lambda_i^{(k)} E_i^{(k)} \right\|_2$
subject to	$U^{(m,n)} \cdot \text{diag}\left(X^{(1)}, .., X^{(N-1)}\right) = \beta^2 \left(T_m, T_n\right) T_m$
	(one equality constraint per fitted swaption)
and	$X^{(k)} \succeq 0,\ \lambda_i^{(k)} \geq 0 \quad i = 1, .., 5;\ k = 1, .., N-1$

which may be expressed in the standard form (A.9) of Section-A.4.3.

Chapter 8

Interpolating Between Nodes

The business need for variable coverages δ_j, the backward construction technique of Section-3.2, together with a limited calibration set, leads to a situation in which the forwards $K(t,T)$ and their corresponding volatilities $\xi(t,T)$ will be known only at a discrete maturity set T_j $(j = 1,..,n)$. To price an instrument depending on an intermediate maturity, for example, a caplet maturing between T_j and T_{j+1}, we therefore need to *interpolate* both volatility and forward.

Our approach is to interpolate on deterministic functions like $\xi(t,T)$, and then use properties of the model to derive interpolations for stochastic variables like $K(t,T)$ and discount functions $B(t,T)$. In contrast, direct interpolation on stochastic variables turns out to be inaccurate and unsatisfactory.

Note that none of the methods described in this chapter are arbitrage-free, though in practice they work fairly accurately. Moreover, they are also inconsistent in that Section-8.3 on interpolating discount factors ought to determine how the forwards in Section-8.1 are interpolated, but they don't. The author suggests consulting Schlogl [113] for a more exacting analysis.

8.1 Interpolating forwards

For $0 \leq t \leq T_0 < T < T_1$ suppose $K(t,T_0)$ and $K(t,T_1)$ are known and $K(t,T)$ is required (for example, to find the intrinsic value of a forward swap whose resets fall between nodes). Start by interpolating the vector volatility function $\xi(t,T)$ on maturity T, defining

$$
\begin{aligned}
\xi(t,T) &= \frac{1}{\delta}\theta(T)\left\{(T_1 - T)\,\xi(t,T_0) + (T - T_0)\,\xi(t,T_1)\right\}, && (8.1)\\
&= \alpha(T)\,\xi(t,T_0) + \beta(T)\,\xi(t,T_1),\\
\textit{with} \quad & \theta(T_0) = 1 = \theta(T_1).
\end{aligned}
$$

This interpolation preserves correlation between forwards at nodepoints, $\theta(\cdot)$ can be chosen to satisfy some auxiliary condition, and both $\alpha(T)$ and $\beta(T)$ are independent of simulation paths and so can be precomputed.

Recalling

$$
\begin{aligned}
& K(t,T_0) = K(0,T_0)\,\mathcal{E}\left\{\int_0^t \xi^*(s,T_0)\,dW_{T_1}(s)\right\} \\
\Rightarrow\quad & \ln\frac{K(t,T_0)}{K(0,T_0)} = \int_0^t \xi^*(s,T_0)\,dW_{T_1}(s) - \frac{1}{2}\int_0^t |\xi(s,T_0)|^2\,ds
\end{aligned}
$$

and identifying the forward measures for $K(t,T)$, $K(t,T_1)$ with that for $K(t,T_0)$

$$
\int_0^t \xi^*(s,T)\,dW_{T_1}(s) = \left\{\begin{array}{c} \alpha(T)\int_0^t \xi^*(s,T_0)\,dW_{T_1}(s) \\ +\beta(T)\int_0^t \xi^*(s,T_1)\,dW_{T_1}(s)\end{array}\right\} \quad so
$$

$$
\begin{aligned}
\ln\frac{K(t,T)}{K(0,T)} &\cong \alpha(T)\ln\frac{K(t,T_0)}{K(0,T_0)} + \beta(T)\ln\frac{K(t,T_1)}{K(0,T_1)} \\
&\quad + \frac{1}{2}\left[\begin{array}{r}\alpha(T)\int_0^t |\xi(s,T_0)|^2\,ds + \beta(T)\int_0^t |\xi(s,T_1)|^2\,ds \\ -\int_0^t |\xi(s,T)|^2\,ds\end{array}\right],
\end{aligned} \tag{8.2}
$$

an interpolation for $K(t,T)$ in terms of $K(t,T_0)$ and $K(t,T_1)$ that contains a convexity term which can be precomputed using approximations like

$$
\int_0^t |\xi(s,T)|^2\,ds = \frac{t}{T}\int_0^T |\xi(s,T)|^2\,ds.
$$

REMARK 8.1 Interpolations that are closer to being arbitrage-free and consistent with the model, can be obtained by distinguishing measures and approximating the corresponding drifts. □

8.2 Dead forwards

For $T_0 < T < T_1$ if $K(T_0,T_0)$ and $K(T,T_1)$ are known and $K(T,T)$ is required (for example, to price a caplet maturing at T), the difficulty is the *dead* forward $K(T_0,T_0)$. To permit sensible approximations while maintaining stochasticity, we *extend the life* of $K(t,T_0)$ to T by writing

$$
\xi(t,T_0) = \xi(T_0,T_0) \quad \text{for} \quad T_0 \le t \le T,
$$

and then compute the *virtual forward* $K(T,T_0)$, given the three forward values $K(T_0,T_0)$, $K(T_0,T_1)$ and $K(T,T_1)$, from the conditional expectation

$$
\frac{K(T,T_0)}{K(T_0,T_0)} = \mathbf{E}_{T_1}\left\{\mathcal{E}\left\{\begin{array}{c}\xi^*(T_0,T_0) \\ \times[W_{T_1}(T) - W_{T_1}(T_0)]\end{array}\right\}\middle|\int_{T_0}^T \xi^*(s,T_1)\,dW_{T_2}(s)\right\},
$$

where $T_2 = T_1 + \delta_1$. Identifying the two measures $\mathbb{P}_{T_1}$ and $\mathbb{P}_{T_2}$ and setting

$$q_0 = |\xi(T_0, T_0)|^2 (T - T_0), \quad q_1 = \int_{T_0}^{T} |\xi(s, T_1)|^2 ds,$$

$$q_{12} = \int_{T_0}^{T} \xi^*(T_0, T_0)\, \xi(s, T_1)\, ds,$$

$$Y = \int_{T_0}^{T} \xi^*(s, T_1)\, dW_{T_1}(s) \cong \ln\left(\frac{K(T, T_1)}{K(T_0, T_1)}\right) + \frac{1}{2} q_1,$$

then $$\{\xi^*(T_0, T_0)[W_{T_1}(T) - W_{T_1}(T_0)] \,|, Y\} \sim \mathbf{N}\left(\frac{q_{12}}{q_1} Y,\ q_0 - \frac{q_{12}^2}{q_1}\right),$$

so $$\frac{K(T, T_0)}{K(T_0, T_0)} \cong \exp\left\{\frac{q_{12}}{q_1} Y + \frac{1}{2}\left(q_0 - \frac{q_{12}^2}{q_1}\right) - \frac{1}{2} q_0\right\},$$

or $$\ln\frac{K(T, T_0)}{K(T_0, T_0)} \cong \frac{q_{12}}{q_1} \ln\left(\frac{K(T, T_1)}{K(T_0, T_1)}\right) + \frac{1}{2} q_{12}\left(1 - \frac{q_{12}}{q_1}\right). \tag{8.3}$$

Combining (8.2) evaluated at $t = T$ and (8.3) gives an interpolation for $K(T, T)$.

8.3 Interpolation of discount factors

Iterating (1.9) relates the zeros $B(t, T_N)$ and $B(t, T_j)$ when $T_j < T_N$

$$B(t, T_N) = \frac{B(t, T_j)}{\prod_{k=j}^{N-1} [1 + \delta K(t, T_k)]}, \quad B(T_j, T_N) = \frac{1}{\prod_{k=j}^{N-1} [1 + \delta K(t, T_k)]}.$$

During a simulation at a time $t = T_j$ on the tenor node T_N, we therefore know exactly the discount factor $B(T_j, T_N)$. The problem arises when discount factors between nodes are required; for example, to make a decision at $t = T \in (T_0, T_1)$ because

$$B(T, T_n) = \frac{B(T, T_1)}{\prod_{j=1}^{n-1} [1 + \delta K(T, T_j)]},$$

requires an expression for $B(T, T_1)$.

The interpolation scheme (8.3) gives $K(T, T)$, so assuming a constant infinitesimal HJM forward rate f on $[T, T + \delta]$ when $t = T$, we have

$$B(T, T + \delta) = \frac{1}{1 + \delta K(T, T)} = \exp[-f\delta] \qquad \Rightarrow$$

$$B(T, T_1) = \exp[-f(T_1 - T)] = \frac{1}{(1 + \delta K(T, T))^{\frac{(T_1 - T)}{\delta}}}.$$

Hence the interpolation scheme for discount factors

$$B(T,T_n) = \frac{1}{(1+\delta K(T,T))^{\frac{(T_1-T)}{\delta}} \prod_{j=1}^{n-1} [1+\delta K(T,T_j)]}.$$

8.4 Consistent volatility

The interpolation can be used to define $\xi(t,T)$ for all T in a fashion *consistent* with the above interpolations on forwards and zeroes. For example, if we ask that caplet volatilities are *linearly interpolated* then

$$\sqrt{\frac{\int_0^T |\xi(s,T)|^2 ds}{T}} = \left\{ \begin{array}{c} \frac{(T_1-T)}{\delta}\sqrt{\frac{\int_0^{T_0}|\xi(s,T_0)|^2 ds}{T_0}} \\ +\frac{(T-T_0)}{\delta}\sqrt{\frac{\int_0^{T_1}|\xi(s,T_1)|^2 ds}{T_1}} \end{array} \right\}.$$

Integrating (8.1) and equating $\int_0^T |\xi(s,T)|^2 ds$ in the two equations

$$\delta^2 \int_0^T |\xi(s,T)|^2 ds = \theta^2(T) \left\{ \begin{array}{c} (T_1-T)^2 \int_0^T |\xi(s,T_0)|^2 ds \\ +2(T-T_0)(T_1-T)\int_0^T \xi^*(s,T_1)\,\xi(s,T_0)\,ds \\ +(T-T_0)^2 \int_0^T |\xi(s,T_1)|^2 ds \end{array} \right\},$$

determines $\theta(T)$ and hence the required interpolation for $\xi(t,T)$.

Chapter 9

Simulation

Two simulation methods are described in this chapter. *Glasserman type methods* [44], [45] avoid bias from drifts by discretizing the SDEs of positive continuous time martingales, and produce accurate results for time steps of the order of a couple of weeks. *Big-step methods* [90] use predictor-corrector techniques to approximate drift and volatility, and step in intervals of years between decision times (like Bermudan exercises). The author strongly recommends Glassermans's book 'Monte Carlo Methods in Financial Engineering' [46] as a reference for this chapter.

9.1 Glasserman type simulation

This approach takes zeros $B(t,T_j)$ discounted by the numeraire $N(t)$ (either $M(t)$ for $\mathbb{P}_0$ or $B(t,T_n)$ for $\mathbb{P}_n$), and for $j = @(t),..,n$ defines martingales $Z(t,T_j)$ and $V(t,T_j)$ under the corresponding measure by

$$
\begin{gathered}
Z(t,T_j) = \frac{B(t,T_j)}{N(t)} > 0 \qquad \Rightarrow \\
H(t,T_j) = K(t,T_j) + a(T_j) = \frac{Z(t,T_j) - \lambda_j Z(t,T_{j+1})}{\delta_j Z(t,T_{j+1})} > 0 \quad (j < n) \\
\textit{in which the} \quad \lambda_j = 1 - \delta_j a(T_j) \quad \textit{satisfy} \quad 0 \le \lambda_j \le 1; \\
V(t,T_j) = Z(t,T_j) - \lambda_j Z(t,T_{j+1}) > 0 \quad (j < n) \ \textit{and} \ V(t,T_n) = Z(t,T_n); \\
Z(t,T_j) = V(t,T_j) + \Pi_j^j V(t,T_{j+1}) ... + \ \Pi_j^{n-1} V(t,T_n), \quad \Pi_j^k = \lambda_j \lambda_{j+1} ... \lambda_k.
\end{gathered}
$$

Discretization of the $Z(t,T_j)$ and $V(t,T_j)$ as martingales can be rigorously enforced and decreases bias by avoiding drift (unlike, for example, simulating $K(t,T_j)$ under $\mathbb{P}_n$). Both $Z(t,T_j)$ and $V(t,T_j)$ should be strictly positive (zeros, numeraires and shifted forwards are strictly positive), which can be ensured by integrating a lognormal SDE with exponential increments.

Making diffusion coefficients Lipschitz in the discrete case without changing the continuous time dynamics turns out to require a simple adjustment.

Introduce $\phi\{x\} = \min\{1, x^+\}$ and rely on Remarks 3.1.1 that

$$0 < h_j(t) = \frac{\delta_j H(t, T_j)}{[1 + \delta_j K(t, T_j)]} = \frac{V(t, T_j)}{Z(t, T_j)} < 1,$$

to rewrite bond volatility differences as

$$b(t, T_j, T_{j+1}) = h_j(t)\,\xi(t, T_j) = \phi\{h_j(t)\}\,\xi(t, T_j) = \phi\left\{\frac{V(t, T_j)}{Z(t, T_j)}\right\}\xi(t, T_j).$$

9.1.1 Under the terminal measure $\mathbb{P}_n$

From (2.4) and (3.3) clearly $Z(t, T_n) = 1$ and SDEs for the $Z(\cdot)$ for $j = @(t), .., n-1$ are

$$\begin{aligned}\frac{dZ(t, T_j)}{Z(t, T_j)} &= b^*(t, T_j, T_n)\, dW_n(t) \\ &= \left[\sum_{\ell=j}^{n-1} \phi\left\{1 - \lambda_\ell \frac{Z(t, T_{\ell+1})}{Z(t, T_\ell)}\right\} \xi^*(t, T_\ell)\right] dW_n(t),\end{aligned}$$

while $V(t, T_n) = 1$ and SDEs for the $V(\cdot)$ for $j = @(t), .., n-1$ are

$$\begin{aligned}\frac{dV(t, T_j)}{V(t, T_j)} &= \left[\xi^*(t, T_j) + \sum_{\ell=j+1}^{n-1} \phi\left\{\frac{V(t, T_\ell)}{Z(t, T_\ell)}\right\} \xi^*(t, T_\ell)\right] dW_n(t), \\ Z(t, T_\ell) &= V(t, T_\ell) + \Pi_\ell^\ell V(t, T_{\ell+1}) ... + \Pi_\ell^{n-1} V(t, T_n),\end{aligned}$$

in which $\sum_{\ell=n}^{n-1}[\cdot] = 0$ when $j = n-1$.

9.1.2 Under the spot measure $\mathbb{P}_0$

From (5.5) and (3.3) SDEs for the $Z(\cdot)$ are $dZ(t, T_@) = 0$ and for $j = @(t) + 1, .., n$

$$\begin{aligned}\frac{dZ(t, T_j)}{Z(t, T_j)} &= -b^*(t, T_@, T_j)\, dW_0(t) \\ &= -\left[\sum_{\ell=@(t)}^{j-1} \phi\left\{\frac{V(t, T_\ell)}{Z(t, T_\ell)}\right\} \xi^*(t, T_\ell)\right] dW_0(t).\end{aligned}$$

while corresponding SDEs for the $V(\cdot)$ for $j = @(t), .., n$ are

$$\begin{aligned}\frac{dV(t, T_j)}{V(t, T_j)} &= \left[\xi^*(t, T_j) - \sum_{\ell=@(t)}^{j} \phi\left\{\frac{V(t, T_\ell)}{Z(t, T_\ell)}\right\} \xi^*(t, T_\ell)\right] dW_0(t), \\ Z(t, T_\ell) &= V(t, T_\ell) + \Pi_\ell^\ell V(t, T_{\ell+1}) ... + \Pi_\ell^{n-1} V(t, T_n).\end{aligned}$$

Note there is some redundancy in these $(n - @(t) + 1)$ SDEs for $V(t, T_j)$ because

$$dZ(t, T_@) = 0 = d\left\{V(t, T_@) + \Pi_\ell^\ell V(t, T_{@+1}) ... + \Pi_\ell^{n-1} V(t, T_n)\right\}.$$

REMARK 9.1 In simulating in the Glasserman framework, bear in mind:
[1] The $V(t, T_j)$ are differences of the $Z(t, T_j)$ and ought to exhibit approximately the same sort of order of magnitude behavior as the forwards $K(t, T_j)$, hence they simulate more accurately than the $Z(t, T_j)$.
[2] In constructing payoff functions use the property that discounted swaps will be linear combinations of the $V(t, T_j)$ because from (2.10)

$$\frac{\text{pSwap}(t)}{N(t)} = \begin{array}{c} Z(t, T_{j0}) - Z(t, T_{jN}) \\ + \sum_{j=j0}^{jN-1} \delta_j \mu_j Z(t, T_{j+1}) - \kappa \sum_{i=i0}^{iM-1} \overline{\delta}_i Z(t, \overline{T}_{i+1}). \end{array} \tag{9.1}$$

[3] Integrate the lognormal SDEs in $V(\cdot)$ with exponential increments putting

$$V_{t+\Delta t} = V_t \exp\left\{(vol\text{-}at\text{-}t)^* \Delta W_t - \frac{1}{2}|(vol\text{-}at\text{-}t)|^2 \Delta t\right\}.$$

[4] Usually a quarterly timestep $\Delta t = \frac{1}{4}$ is quite accurate (trial it on caps), but always check results with a two-weekly timestep of $\Delta t = \frac{1}{24}$. □

9.2 Big-step simulation

From (5.5.1) and (5.5.7) the shifted forward $H(t, T_j)$ is either

$$\frac{H(t, T_j)}{H(0, T_j)} = \mathcal{E}\left\{\begin{array}{c} -\int_0^t \left[\sum_{\ell=j+1}^{n-1} h_\ell(s) \xi^*(s, T_\ell)\right] \xi(s, T_j) \, ds \\ + \int_0^t \xi^*(s, T_j) \, dW_n(s) \end{array}\right\} \tag{9.2}$$

$$or \quad \frac{H(t, T_j)}{H(0, T_j)} = \mathcal{E}\left\{\begin{array}{c} \int_0^t \left[\sum_{\ell=@(s)}^{j} h_\ell(s) \xi^*(s, T_\ell)\right] \xi(s, T_j) \, ds \\ + \int_0^t \xi^*(s, T_j) \, dW_0(s) \end{array}\right\} \tag{9.3}$$

under the terminal $\mathbb{P}_n$ or the spot $\mathbb{P}_0$ measure respectively. In either case, to simulate $H(t, T_j)$ from $H(0, T_j)$ in one step (that is, by *big-step* simulation) we need to separately produce expressions for the drift and volatility terms in (9.2) and (9.3). Simulation of subsequent steps then follows the same pattern, because the model is Markov in the $H(t, T_j)$.

9.2.1 Volatility approximation

To jointly produce the normally distributed components of a vector

$$X = (X_j) = \left(\int_0^t \xi^*(s, T_j) \, dW(s)\right)$$

construct a new high-dimensional model with a volatility function piecewise constant over the simulation interval $[0, t]$ that has the same finite dimensional distributions as the original model. Do that by finding the square root Γ of the covariance matrix

$$q = \mathbf{E}XX^T = \mathbf{E}\left(X_{j1}\ X_{j2}\right) = \frac{1}{t}\left(\int_0^t \xi^*\left(s, T_{j1}\right) \xi\left(s, T_{j2}\right) ds\right) = \Gamma\Gamma^T,$$

and defining the volatility function $\widetilde{\xi}\left(\cdot\right)$ of the new model to be

$$\widetilde{\xi}\left(s, T_j\right) = \Gamma_{j,\cdot} \qquad s \in [0, t].$$

The simulation step $X = \sqrt{t}\Gamma\varepsilon$, where ε is a vector of IID standard normal random variables, then produces identical covariance.

9.2.2 Drift approximation

The drift terms in both (9.2) and (9.3) involves integrals like

$$D\left(t, T_\ell, T_j\right) = \int_0^t h_\ell\left(s\right) \xi^*\left(s, T_\ell\right) \xi\left(s, T_j\right) ds,$$

$$h_\ell\left(s\right) v = \frac{\delta'_\ell H\left(s, T_\ell\right)}{1 + \delta'_\ell H\left(s, T_\ell\right)}, \qquad \delta'_\ell = \frac{\delta_\ell}{1 - \delta_\ell a_\ell},$$

in which $h_\ell\left(s\right)$ is a $\mathbb{P}_\ell$-martingale lying in $(0, 1)$ that must be approximated knowing either $h_\ell\left(t\right)$ or $H\left(t, T_\ell\right)$ in order to *big-step* over $[0, t]$.

REMARK 9.2 It turns out that the 'initial approximation' $h_\ell\left(s\right) = h_\ell\left(0\right)$ performs quite well, so clearly any reasonable approximation incorporating $h_\ell\left(t\right)$ will perform better. Hence the following approach (similar to that of Pelsser et al [90]) using fairly robust approximations in various places. □

For $0 < s < t \leq T < T_1 = T + \delta$, appropriate approximations for a typical shifted forward $H\left(t, T\right)$, are

$$\frac{\delta' H\left(s, T\right)}{1 + \delta' H\left(s, T\right)} \cong \mathbf{E}_T\left\{\left.\frac{\delta' H\left(s, T\right)}{1 + \delta' H\left(s, T\right)}\right| H\left(t, T\right)\right\}$$

$$= \mathbf{E}_{T_1}\left\{\left.\frac{\delta' H\left(s, T\right)}{1 + \delta' H\left(s, T\right)}\right| H\left(t, T\right)\right\} \cong \frac{\delta' \mathbf{E}_{T_1}\left\{H\left(s, T\right) | H\left(t, T\right)\right\}}{1 + \delta' \mathbf{E}_{T_1}\left\{H\left(s, T\right) | H\left(t, T\right)\right\}},$$

$$M\left(s, T\right) = \int_0^s \xi^*\left(u, T\right) dW_{T_1}\left(u\right), \qquad q\left(s, T\right) = \int_0^s \left|\xi\left(u, T\right)\right|^2 du,$$

$$H\left(s, T\right) = H\left(0, T\right) \exp\left\{M\left(s, T\right) - \tfrac{1}{2} q\left(s, T\right)\right\}.$$

Noting the random variable $\{M(s,T)|\, M(t,T)\}$ is normally distributed

$$\{M(s,T)\,|\,M(t,T)\} \sim \mathbf{N}\left\{\tfrac{q(s,T)}{q(t,T)}\; M(t,T),\; \tfrac{q(s,T)}{q(t,T)}\left[q(t,T)-q(s,T)\right]\right\};$$
$$\mathbf{E}_{T_1}\{H(s,T)\,|\,H(t,T)\} = H(0,T)\exp\left\{\tfrac{q(s,T)}{q(t,T)}M(t,T) - \tfrac{1}{2}\tfrac{q(s,T)}{q(t,T)}q(s,T)\right\},$$
$$\cong H(0,T)^{1-\frac{q(s,T)}{q(t,T)}}\; H(t,T)^{\frac{q(s,T)}{q(t,T)}}$$

and therefore the drift can be approximated by

$$D(t,T_\ell,T_j) \cong \int_0^t \frac{\delta' H(0,T_\ell)^{1-\frac{q(s,T_\ell)}{q(t,T_\ell)}}\; H(t,T_\ell)^{\frac{q(s,T_\ell)}{q(t,T_\ell)}}}{1+\delta' H(0,T_\ell)^{1-\frac{q(s,T_\ell)}{q(t,T_\ell)}}\; H(t,T_\ell)^{\frac{q(s,T_\ell)}{q(t,T_\ell)}}}\xi^*(s,T_\ell)\,\xi(s,T_j)\,ds,$$
$$= \frac{\ln\frac{1+\delta' H(t,T_\ell)}{1+\delta' H(0,T_\ell)}}{\ln\frac{H(t,T_\ell)}{H(0,T_\ell)}}q(t,T_\ell) \quad \mathit{if} \quad \xi(s,T_j) = \xi(s,T_\ell) \quad \mathit{for} \quad s\in[0,t].$$

So if the following ratio is independent of s

$$\frac{\xi^*(s,T_\ell)\,\xi(s,T_j)}{|\xi(s,T_\ell)|^2} \quad\Rightarrow\quad \xi^*(s,T_\ell)\,\xi(s,T_j) = \mathrm{fn}(T_\ell,T_j)\,|\xi(s,T_\ell)|^2$$

as it is, for example, when the multi-factor volatility vector $\xi(t,T)$ is either constant or one-factor separable $\xi(t,T)=\chi(t)\,\phi(T)$, then the drift over the simulation step can be approximated by the computationally efficient

$$D(t,T_\ell,T_j) \cong \frac{\ln\frac{1+\delta' H(t,T_\ell)}{1+\delta' H(0,T_\ell)}}{\ln\frac{H(t,T_\ell)}{H(0,T_\ell)}}\int_0^t \xi^*(u,T_\ell)\,\xi(u,T_j)\,du, \tag{9.4}$$
$$= \frac{\ln\frac{1+\delta' H(t,T_\ell)}{1+\delta' H(0,T_\ell)}}{\ln\frac{H(t,T_\ell)}{H(0,T_\ell)}}\; q_{\ell,j}\; t.$$

REMARK 9.3 An alternative view of (9.4) is that it can be obtained using an *exponential bridge* with $\xi(s,T_\ell)$ and $\xi(s,T_j)$ constant on $s\in[0,\ t]$

$$H(s,T_\ell) = H(0,T_\ell)^{1-\frac{s}{t}}\; H(t,T_\ell)^{\frac{s}{t}}.$$

As a further possibility, that suggests

$$h_\ell(s) = \left(1-\frac{s}{t}\right)h_\ell(0) + \frac{s}{t}h_\ell(t)$$

which is a *linear bridge*. ☐

9.2.3 Big-stepping under the terminal measure $\mathbb{P}_n$

Apply the approximations obtained in Section-9.2.1 and Section-9.2.2 to equation (9.2). Work backwards, and sequentially compute the shifted forwards $H(t, T_{n-1})$,..., $H(t, T_{@})$; at each step the approximations can be directly applied.

9.2.4 Big-stepping under a *tailored* spot measure $\overline{\mathbb{P}}_0$

The biggest time steps that can be taken using (9.3) are clearly the coverages δ_j, because the summation in the drift in (9.3) is from $\ell = @(s)$. To big-step between nodes $\overline{T}_i$ $(i = 1, .., M)$, which (abusing the swap notation of Section 2.2) we assume are decision times for some product (for example, a Bermudan), introduce a new *tailored spot measure* $\overline{\mathbb{P}}_0$ specific to the product by using as numeraire $\overline{M}(t)$ the roll-up of zero coupons between the decision nodes $\overline{T}_i$

$$\overline{M}(t) = B\left(t, \overline{T}_{\overline{@}}\right) \prod_{\ell=0}^{\overline{@}-1} \frac{1}{B\left(\overline{T}_\ell, \overline{T}_{\ell+1}\right)}, \qquad \overline{@}(t) = \inf\left\{\ell \in \mathbb{Z}^+ : t \leq \overline{T}_\ell\right\}.$$

Similarly from Section-5.3 and using $\overline{bar}$ notation in the obvious way, if

$$\begin{aligned} Z(t, T_j) &= \frac{B(t, T_j)}{\overline{M}(t)}, \quad t \leq T_j, \qquad \textit{then} \\ \frac{dZ(t, T_j)}{Z(t, T_j)} &= -\left[b^*\left(t, T_{\overline{@}}\right) - b^*(t, T_j)\right]\left\{dW(t) - b\left(t, T_{\overline{@}}\right)dt\right\}, \\ &= -b^*\left(t, T_{\overline{@}}, T_j\right) d\overline{W}_0(t), \qquad \textit{where} \\ d\overline{W}_0(t) &= dW(t) - b\left(t, T_{\overline{@}}\right)dt \qquad \textit{which implies} \\ dW_{j+1}(t) &= d\overline{W}_0(t) + b\left(t, T_{\overline{@}}, T_{j+1}\right)dt. \end{aligned}$$

Hence under $\overline{\mathbb{P}}_0$ an expression for $H(t, T_j)$ for $j = \overline{@}(t), .., n-1$, is

$$\frac{H(t, T_j)}{H(0, T_j)} = \mathcal{E}\left\{\begin{array}{c} \int_0^t \left[\sum_{\ell=\overline{@}(s)}^{j} h_\ell(s)\, \xi^*(s, T_\ell)\right] \xi(s, T_j)\, ds \\ + \int_0^t \xi^*(s, T_j)\, d\overline{W}_0(s) \end{array}\right\}, \tag{9.5}$$

and because the summation is now from $\ell = \overline{@}(s)$, the difficulties that occurred in big-stepping under the spot measure $\mathbb{P}_0$ disappear.

A further slight difficulty is that successive $H(t, T_j)$ for $j = \overline{@}(t), .., n-1$ occur on both sides of (9.5). Tackle by *cycling*: find a value for $H(t, T_j)$ by setting $h_j(s) = h_j(0)$, put that value into the approximations of Section-9.2.1 and Section-9.2.2 to compute a new $H(t, T_j)$, and then repeat.

Finally, payoffs are of course present valued using the tailored spot measure $\overline{\mathbb{P}}_0$. For $t \leq T^* \leq T < T_n$ the value $X(t)$ of a cashflow $X(T^*)$ determined at T^* and made at T, is therefore given by

$$\frac{X(t)}{\overline{M}(t)} = \mathbf{E}_{\overline{\mathbb{P}}_0}\left\{ \left. \frac{X(T^*)}{\overline{M}(T)} \right| \mathcal{F}_t \right\} = \mathbf{E}_{\overline{\mathbb{P}}_0}\left\{ \frac{B(T^*, T)\, X(T^*)}{\overline{M}(T^*)} \right\}.$$

Chapter 10

Timeslicers

If the shifted BGM volatility function is *separable* $\xi(t,T) = \chi(t)\ \phi(T)$ so

$$\int_0^t \xi^*(t,T)\,dW(t) = \phi^*(T)\int_0^t \chi(t)\,dW(t) = \phi^*(T)\,M(t)$$

then the *driver* in the model becomes the Gaussian martingale $M(t)$ which (apropos of drifts) permits valuation by PDEs, lattices, trees or *timeslicers.* But getting a multi-dimensional homogenous correlation function $c(T-t)$ into this framework is problematical (the last section in this chapter shows how it might be approached), so we concentrate on the one-factor case, which usually produces good theta and vega approximations for exotics, and which can also be used to debug code in corresponding simulation routines.

Consider a simple Black-Scholes (BS) model (the notation is obvious)

$$\beta_t = \exp(rt), \quad S_t = S_0\mathcal{E}(rt+\sigma W_t) = S(t,X_t), \quad X_t = \sigma W_t \sim \mathbf{N}\left(0,\sigma^2 t\right),$$

driven by the Gaussian martingale X_t. If $f_t = f(t,X_t)$ is the time t price of some derivative, then for $0 < T_L < T_R$ (lying on floating-side nodes)

$$\frac{f(T_L,X_{T_L})}{\beta_{T_L}} = \mathbf{E}_0\left\{\left.\frac{f(T_R,X_{T_R})}{\beta_{T_R}}\right|\mathcal{F}_{T_L}\right\}.$$

Timeslicing consists of computing these conditional expectations backwards from maturity to root across consecutive timeslices $T_R \to T_L$ in a semi-analytic fashion (integrating cubics against normal densities) using the fact that X_{T_R} conditioned on X_{T_L} is also normally distributed (see Section-A.2.1):

$$\begin{aligned}(X_{T_R}|\,X_{T_L}) &\sim \mathbf{N}\left(X_{T_L},\sigma^2[T_R-T_L]\right) \qquad \Rightarrow\\ f(T_L,X_{T_L}) &= \int_{-\infty}^{\infty} e^{-r[T_R-T_L]}f\left(T_R,X_{T_L}+\sigma\sqrt{T_R-T_L}x\right)\mathbf{N}_1(x)\,dx.\end{aligned}$$

To evaluate integrals across timeslices $T_R \to T_L$ for each X_{T_L}-node:

Step-1 Represent the random variables X_{T_L} and X_{T_R} by 50-100 nodes ranging to 5-6 standard deviations of X_t on respective timeslices,

Step-2 Values of $f(T_R,X_{T_R})$ will be attached to the X_{T_R}-nodes on timeslice-T_R; cubic spline them to produce a continuous representation,

Step-3 Analytically evaluate the integrals, which are cubics against Gaussian densities, to find $f(T_L,X_{T_L})$ at each.X_{T_L}-node on timeslice-T_L.

REMARK 10.1 To calibrate judgement about accuracy, node spacing and speed, value a BS call, whose price is known, back through about 20 timeslices. Depending on node density 4 to 6 figures of accuracy are possible. □

REMARK 10.2 The timeslice approach is flexible. For example, if the timeslices are the resets of a Bermudan swaption, node values at timeslice-T will be the *continuation value* for some X_T and the *intrinsic value* for other X_T. □

10.1 Terminal measure timeslicer

With a one-factor separable volatility function $\xi(t,T) = \chi(t)\,\phi(T)$, from (9.2) the shifted forwards under the terminal measure $\mathbb{P}_n$ take the form

$$\frac{H(t,T_j)}{H(0,T_j)} = \exp\left\{\begin{array}{c}\phi(T_j)\,M_n(t) - \frac{1}{2}\phi^2(T_j)\,q(t) \\ -\phi(T_j)\sum_{\ell=j+1}^{n-1}\phi(T_\ell)\int_0^t h_\ell(s)\,dq(s)\end{array}\right\},$$

$$M_n(t) = \int_0^t \chi(s)\,dW_n(s), \quad q(t) = \langle M_n\rangle(t) = \int_0^t \chi^2(s)\,ds, \quad j = L,..,n-1$$

and are essentially driven by the Gaussian martingale $M_n(t)$, whose quadratic variation is $q(t)$. Once nodes on say timeslice-T_L are distributed to cover 5 to 6 standard deviations of $M_n(T_L)$, values of $H(T_L,T_j)$ can be assigned to those nodes using the *drift approximation* (9.4)

$$\int_0^{T_L} h_l(s)\,dq(s) \cong \frac{\ln\frac{1+\delta'_\ell H(T_L,T_j)}{1+\delta'_\ell H(0,T_j)}}{\ln\frac{H(T_L,T_j)}{H(0,T_j)}}\,q(T_L), \quad \delta'_\ell = \frac{\delta_\ell}{1-\delta_\ell a_\ell} \tag{10.1}$$

by starting with $H(T_L,T_{n-1})$ and working backwards to specify $H(T_L,T_j)$ for $j = n-1,..,L$. With the forwards $H(T_L,T_j)$ specified the values of other quantities that depend on them, like swaps, swaprates, and the numeraire bond $B(T_L,T_n)$, can also be assigned to nodes on this timeslice-T_L, for example,

$$B(T_L,T_n) = F_{T_L}(T_L,T_n) = \prod_{j=j}^{n-1}\frac{1}{1+\delta_j\left[H(T_L,T_j) - a(T_j)\right]}.$$

Because (apropos of drift) the $\mathbb{P}_n$-numeraire $B(t,T_n)$ is not path dependent (in contrast to the spot numeraires) one can *load* onto the timeslicer either actual or discounted values and then work consistently with that choice; that is, the discounted by $B(t, T_n)$ value $\underline{f}(t)$ of an asset $f(t)$ is a $\mathbb{P}_n$-martingale

$$\underline{f}(t) = \frac{f(t)}{B(t, T_n)} \quad \Rightarrow \quad \underline{f}(T_L) = \mathbf{E}_n \left\{ \underline{f}(T_R) \middle| \mathcal{F}_{T_L} \right\} \cong \mathbf{E}_n \left\{ \underline{f}(T_R) \middle| M_n(T_L) \right\},$$

from our assumption that the model is approximately Markov in $M_n(T_L)$. Because values of $f(t)$ and $B(t, T_n)$ can be assigned to nodes, there is no problem loading the discounted value $\underline{f}(\cdot)$ itself, which can then be timesliced back to the root, where the PV can be recovered as $f(0) = B(0, T_n)\underline{f}(0)$.

To step down timeslices $T_R \to T_L$ the procedure is:

Step-1 Discounted values $\underline{f}(T_R, M_n(T_R))$ will already be attached to the nodes representing $M_n(T_R)$ on timeslice-T_R, so spline them to get a continuous cubic representation.

Step-2 Assign nodes to timeslice-T_L to represent the Gaussian random variable $M_n(T_L)$, and then work out via $H(T_L, T_j)$ the *discounted intrinsic value* $\underline{i}[M_n(T_L)]$ of relevant instruments on each node $M_n(T_L)$, for example, the underlying swap in the case of a Bermudan (values would also be initially attached to the furthest, that is, the very first timeslice in this same fashion).

Step-3 With the aid of the formulae in Section-10.4 and the fact that

$$\{ M_n(T_R) | M_n(T_L) \} \sim \mathbf{N}(M_n(T_L), \; q(T_R) - q(T_L)),$$

compute *discounted continuation values* $\underline{c}[M_n(T_L)]$ at each node $M_n(T_L)$

$$\underline{c}[M_n(T_L)] = \mathbf{E}_n \left\{ \underline{f}(T_R, M_n(T_R)) \middle| M_n(T_L) \right\}.$$

Step-4 From the intrinsic and continuation values, compute timeslice-T_L values at each node $M_n(T_L)$ on timeslice-T_L; for example, for Bermudans

$$\underline{f}(T_L, M_n(T_L)) = \max \{ \underline{c}[M_n(T_L)], \; \underline{i}[M_n(T_L)] \}.$$

Step-5 Go to Step 1, iterate and get $f(0) = B(0, T_n) \; \underline{f}(0)$ at the root.

10.2 Intermediate measure timeslicer

Here the modus operandi is to attach nodes to the timeslicers under their respective forward measure, that is, nodes $M_L(T_L)$ to timeslice-T_L under $\mathbb{P}_{T_L}$ and nodes $M_R(T_R)$ to timeslice-T_R under $\mathbb{P}_{T_R}$, and then step down timeslices $T_R \to T_L$ computing actual values (that is, not discounted values) under the $\mathbb{P}_{T_R}$-forward measures while making some necessary drift adjustments.

Assuming actual (that is, not discounted) values $f(T_R, M_R(T_R))$ are hung on timeslice-T_R, then actual values on timeslice-T_L will be given by

$$f(T_L, M_L(T_L)) = B(T_L, T_R) \mathbf{E}_{T_R} \{ f(T_R, M_R(T_R)) | \mathcal{F}_{T_L} \}.$$

With separable volatility $\xi(t,T) = \chi(t)\,\phi(T)$, from (3.3) some relevant Brownian motions, forwards and driving martingales under $\mathbb{P}_L$ are:

$$dW_{T_j}(t) = dW_{T_L}(t) + \sum_{\ell=L}^{j-1} h_\ell(t)\,\chi(t)\,\phi(T_\ell)\,dt, \quad j > L \tag{10.2}$$

$$\frac{H(t,T_j)}{H(0,T_j)} = \exp\left\{ \begin{array}{c} \phi(T_j)\,M_L(t) - \frac{1}{2}\phi^2(T_j)\,q(t) \\ +\phi(T_j)\sum_{\ell=L}^{j-1}\phi(T_\ell)\int_0^t h_\ell(s)\,\chi^2(s)\,ds \end{array} \right\},$$

$$M_L(t) = \int_0^t \chi(s)\,dW_L(s), \quad q(t) = \langle M_L\rangle(t) = \int_0^t \chi^2(s)\,ds,$$

allowing, after a drift approximation like (10.1), the time-T_L intrinsic values of relevant instruments including $B(T_L,T_R) = F_{T_L}(T_L,T_R)$ to be hung on the nodes $M_L(T_L)$ of timeslice-T_L. All that remains is to make a drift adjustment in the conditional expectation, that is, find the distribution under $\mathbb{P}_{T_R}$ of $M_R(T_R)$ conditional on $M_L(T_L)$, which from (10.2) are connected by

$$M_R(T_L) = M_L(T_L) + \sum_{\ell=L}^{R-1} \phi(T_\ell)\int_0^{T_L} h_\ell(s)\,\chi^2(s)\,ds = M_L(T_L) + D(T_L),$$

$$M_R(T_R) = \{M_L(T_L) + D(T_L)\}\,\mathcal{E}\left\{\int_{T_L}^{T_R} \chi(s)\,dW_R(s)\right\},$$

$$\Rightarrow \quad \{M_R(T_R)|\,M_L(T_L)\} \sim \mathbf{N}\left(M_L(T_L) + D(T_L),\; q(T_R) - q(T_L)\right),$$

because from (10.1), $D(T_L)$ is determined solely by $M_L(T_L)$. The procedure for stepping down nodes $T_R \to T_L$ is then similar to the terminal case Section-10.1, except that at Step-2 the drift term $D(T_L)$ must also be attached to the nodes $M_L(T_L)$ ready to use in the conditional expectation of Step-3.

10.3 A spot measure timeslicer is problematical

Unfortunately constructing a timeslicer under either the spot measure $\mathbb{P}_0$ or the *tailored* spot measure $\overline{\mathbb{P}}_0$ is problematical due to the path dependent nature of the spot numeraires, which from (5.4) and Section-9.2.4 are respectively

$$M(t) = B(t,T_@)\prod_{j=0}^{@-1}\frac{1}{B(T_j,T_{j+1})} \quad \textit{and} \quad \overline{M}(t) = B\left(t,\overline{T}_{\overline{@}}\right)\prod_{j=0}^{\overline{@}-1}\frac{1}{B\left(\overline{T}_j,\overline{T}_{j+1}\right)}.$$

In *big-step* simulation the problem is circumvented by using the *tailored* spot measure $\overline{\mathbb{P}}_0$ and conditioning on information at the beginning of each simulation step. But here, from (9.5), the shifted forwards will take the form

$$\frac{H(t,T_j)}{H(0,T_j)} = \exp\left\{ \begin{array}{c} \phi(T_j)\,M_0(t) - \frac{1}{2}\phi^2(T_j)\,q(t) \\ +\phi(T_j)\int_0^t \sum_{\ell=\overline{@}(s)}^{j} \phi(T_\ell)\,h_\ell(s)\,\chi^2(s)\,ds \end{array} \right\},$$

$$M_0(t) = \int_0^t \chi(s)\,d\overline{W}_0(s), \quad q(t) = \langle M_0\rangle(t) = \int_0^t \chi^2(s)\,ds, \quad j \geq \overline{@}(t),$$

which involves integrating the drift from time zero, rather than from the previous timestep. Hence in general, while we can approximate each $h_k(s)$ as a function of s and its initial and final values, the drift as a whole

$$\phi(T_j)\int_0^t \sum_{\ell=\overline{@}(s)}^{j} \phi(T_\ell)\,h_\ell(s)\,\chi^2(s)\,ds \tag{10.3}$$

cannot be reasonably approximated in terms of a driver $M_0(T)$ at some timeslice-T due to the limit $\overline{@}(s)$ in the summation. In particular, this path dependency prevents nodes being loaded with either initial or intrinsic values. Approximating $h_k(s)$ by its initial value $h_k(0)$ works, but leads to inaccuracies at distant timeslices and in out-of-the-money options.

REMARK 10.3 The spot measure jumps through successive forward measures, see Section-5.3, so if an intermediate measure timeslicer works, why not a spot one? The respective conditional distributions of the driving martingales

$$\{M_R(T_R)|\,M_L(T_L)\} \sim \mathbf{N}(M_L(T_L) + D(T_L),\ q(T_R) - q(T_L)) \quad \textit{under } \mathbb{P}_{T_R},$$
$$\textit{and} \quad \{M(T_R)|\,M(T_L)\} \sim \mathbf{N}(M(T_L),\ q(T_R) - q(T_L)) \quad \textit{under } \overline{\mathbb{P}}_0$$

show there is a difference. The point is that if we want the jump from $\mathbb{P}_{T_R}$ back to $\mathbb{P}_{T_L}$ to be regarded as under $\overline{\mathbb{P}}_0$ then appropriate drifts must be used. For $\overline{\mathbb{P}}_0$ that involves the path dependent term (10.3). □

10.4 Some technical points

10.4.1 Node placement

Density and spacing of nodes will determine speed and accuracy, and two alternatives present themselves: we could distribute nodes non-uniformly so that they are denser where the driver is most likely to be or where sensitive decisions have to be made (for example, close to a barrier); alternatively, we could assume all nodes on all timeslices are equally spaced, and then speed calculation by optimizing code with lookup tables.

Equal spacing is easier to program and optimize for speed, but creates nodes where the driver has low probability of going. Distributing nodes where unconditional densities are highest (for example, as in the Gauss-Hermite integration routine) leads to inaccuracies in splining due to big gaps where the density is low. Whichever method is used an accuracy of five to six decimal places is an appropriate target to aim for.

Whether or not we can expect to use just one lookup table for the whole valuation process, depends largely on whether or not the conditional variances change with timeslice. Because in shifted BGM that is the case for any parameterization other than

$$\xi(t,T) = \phi(T) \quad or \quad \chi(t) = 1,$$

we accept finding a new table for each timeslice, and place nodes at each timeslice in any way that makes sense.

10.4.2 Cubics against Gaussian density

Integrals of cubics (or polynomials in general) against the normal density function can be found by expressing them as the sum of integrals of the form

$$I_\nu(\delta,\beta,a) = \int_0^\delta \frac{u^\nu}{\sqrt{2\pi\beta^2}} \exp\left(-\frac{1}{2}\frac{[u-a]^2}{\beta^2}\right) du \Rightarrow I_\nu = a\, I_{\nu-1} + \beta^2 \frac{\partial I_{\nu-1}}{\partial a}$$

by differentiating $I_\nu(\delta,\beta,a)$ with respect to a under the integral sign. The first few partial derivatives of $I_0(\delta,\beta,a)$ with respect to a are known

$$I_0 = \mathbf{N}\left(\frac{\delta-a}{\beta}\right) - \mathbf{N}\left(-\frac{a}{\beta}\right), \quad \beta I_0' = -\mathbf{N}'\left(\frac{\delta-a}{\beta}\right) + \mathbf{N}'\left(-\frac{a}{\beta}\right),$$
$$\beta^2 I_0'' = \mathbf{N}''\left(\frac{\delta-a}{\beta}\right) - \mathbf{N}''\left(-\frac{a}{\beta}\right), \quad \beta^3 I_0''' = -\mathbf{N}'''\left(\frac{\delta-a}{\beta}\right) + \mathbf{N}'''\left(-\frac{a}{\beta}\right),$$

while from the recurrence relationship

$$I_1 = a\, I_0 + \beta^2 I_0', \qquad I_2 = \left(a^2+\beta^2\right) I_0 + 2a\beta^2 I_0' + \beta^4 I_0'',$$
$$I_3 = a\left(a^2+3\beta^2\right) I_0 + 3\beta^2\left(a^2+\beta^2\right) I_0' + 3a\beta^4 I_0'' + \beta^6 I_0'''.$$

10.4.3 Splining the integrand

Given a set of m equally spaced points

$$\{(x_j, z_j) : z_j = z(x_j), \quad (x_{j+1}-x_j) = \delta \quad j = 1,2,..,m\}$$

(with spacing δ) a cubic spline interpolation for $z = z(x)$ on the interval $x \in [x_j, x_{j+1}]$ only, can be expressed, setting $u = x - x_j$, in the form

$$z(x) = z^{(j)}(u) = z_j + \left[\frac{z_{j+1}-z_j}{\delta} - \frac{\delta}{6} z_{j+1}'' - \frac{\delta}{3} z_j''\right] u + \frac{z_j''}{2} u^2 + \frac{\left[z_{j+1}''-z_j''\right]}{6\delta} u^3.$$

Hence if $X \sim \mathbf{N}(\mu, \sigma^2)$ and we know $z(X)$ for discreet values x_j of X, then

$$\mathbf{E}(z(X)) = \int_{-\infty}^{\infty} \frac{z(x)}{\sqrt{2\pi\sigma^2}} \exp\left\{-\frac{1}{2}\frac{[x-\mu]^2}{\sigma^2}\right\} dx$$

$$\cong \sum_j \int_0^{\delta} \frac{z^{(j)}(u)}{\sqrt{2\pi\sigma^2}} \exp\left\{-\frac{1}{2}\frac{[u-(\mu-x_j)]^2}{\sigma^2}\right\} du,$$

which can be evaluated using the results of Section-10.4.2.

10.4.4 Alternative spline

Instead of splining the integrand as in Section-10.4.3 above, one could spline the product of integrand and density and integrate that. A cubic spline interpolation through m points

$$\{(x_j, z_j) : z_j = z(x_j) \quad j = 1, 2, .., m\}$$

can be expressed, see Press et al [98], in the form

$$z = Az_j + Bz_{j+1} + Cz_j'' + Dz_{j+1}'',$$

$$A = \frac{1}{\delta_j}(x_{j+1} - x), \quad B = 1 - A = \frac{1}{\delta_j}(x - x_j),$$

$$C = \frac{1}{6}(A^3 - A)\delta_j^2, \quad D = \frac{1}{6}(B^3 - B)\delta_j^2, \quad \delta_j = (x_{j+1} - x_j).$$

The integral of the splined function is then

$$\int_{x_1}^{x_m} z(x)\, dx = \sum_{j=1}^{m} \left\{\frac{1}{2}\delta_j (z_j + z_{j+1}) - \frac{1}{24}\delta_j^3 (z_j'' + z_{j+1}'')\right\}.$$

10.5 Two-dimensional timeslicer

The problem is that the homogenous correlation part of a general volatility function is basically incompatible with the separable volatility structure needed to construct a timeslicer. Only exponential functions, which are both homogenous and separable, because $\exp(T-t) = \exp(T)\exp(-t)$, seem to offer a solution

Start with the homogeneous part of a general volatility function

$$\xi(t, T) = \chi(t)\ \phi(T)\ \psi(T - t)\ c(T - t),$$

set $x = T - t$ and bestfit its first two components with exponential functions

$$\psi(x)\,c(x) \cong \exp(-\lambda x)\,\mathbf{a} + \exp(-\mu x)\,\mathbf{b}, \quad \mathbf{a} = \begin{pmatrix} a_1 \\ a_2 \end{pmatrix} \quad \mathbf{b} = \begin{pmatrix} b_1 \\ b_2 \end{pmatrix}.$$

With six parameters $\lambda, \mu, a_1, a_2, b_1, b_2$ a not unreasonable fit is possible. Now recalibrate using the separable multi-factor method of Section-7.5 to obtain

$$\xi(t,T) = \psi(t)\,\phi(T)\,\{\exp(-\lambda[T-t])\,\mathbf{a} + \exp(-\mu[T-t])\,\mathbf{b}\}$$

as the new volatility function for our two-dimensional timeslicer. Apropos of drifts, under say the terminal measure $\mathbb{P}_n$ the driving martingale will be

$$\begin{aligned}
&\int_0^t \psi(t)\,\phi(T)\,[\exp(-\lambda[T-t])\,\mathbf{a} + \exp(-\mu[T-t])\,\mathbf{b}]^*\,dW_n(t), \\
&= |\mathbf{a}|\,\phi(T)\exp(-\lambda T)\int_0^t \psi(t)\exp(\lambda t)\,\frac{\mathbf{a}^*}{|\mathbf{a}|}dW_n(t) \\
&\quad + |\mathbf{b}|\,\phi(T)\exp(-\mu T)\int_0^t \psi(t)\exp(\mu t)\,\frac{\mathbf{b}^*}{|\mathbf{b}|}dW_n(t), \\
&= |\mathbf{a}|\,\phi(T)\exp(-\lambda T)\,M_n^a(t) + |\mathbf{b}|\,\phi(T)\exp(-\mu T)\,M_n^b(t),
\end{aligned}$$

a linear combination of two jointly normally distributed random variables

$$M_n^a(t) = \int_0^t \psi(t)\exp(\lambda t)\,dW_a(t), \qquad M_n^b(t) = \int_0^t \psi(t)\exp(\mu t)\,dW_b(t),$$

that have zero mean, and variances, covariance and correlation given by

$$\operatorname{var} M_n^a(t) = \int_0^t \psi^2(s)\exp(2\lambda s)\,ds, \qquad \operatorname{var} M_n^b(t) = \int_0^t \psi^2(s)\exp(2\mu s)\,ds,$$

$$\operatorname{cov}\left(M_n^a(t),\ M_n^b(t)\right) = \frac{\mathbf{a}\cdot\mathbf{b}}{|\mathbf{a}|\,|\mathbf{b}|}\int_0^t \psi^2(s)\exp([\lambda+\mu]\,s)\,ds,$$

$$\rho(t) = \frac{\operatorname{cov}\left(M_n^a(t),\ M_n^b(t)\right)}{\sqrt{\operatorname{var} M_n^a(t)}\sqrt{\operatorname{var} M_n^b(t)}}.$$

A *correlated* two-dimensional timeslicer carrying both $M_n^a(t)$ and $M_n^b(t)$ can now be constructed in a similar way to the one-dimensional equivalent.

Chapter 11

Pathwise Deltas

The most important risk measures are *deltas* (sensitivity to movements of the underlying yield curve) and *vegas* (sensitivity to changes in the implied volatilities of the instruments to which the model is calibrated); in this chapter we show how to compute pathwise deltas along the lines of Glasserman and Zhao [44].

Denote by $\mathbb{P}$ (numeraire N_t expectation $\mathbf{E}$) whichever measure, spot $\mathbb{P}_0$ or terminal $\mathbb{P}_n$, we are using. Let C_0 be the present value of the discounted payoff stream 'payoff $C(\cdot)$', and $D_\ell = \frac{\partial}{\partial K(0,T_\ell)}$ denote partial differentiation with respect to the initial value $K(0,T_\ell)$ of the forward $K(t,T_\ell)$. Then

$$C_0 = N_0 \, \mathbf{E} \, \text{payoff}\, C \quad \Rightarrow \quad D_\ell C_0 = D_\ell \left(N_0 \frac{C_0}{N_0} \right) \qquad \textit{giving}$$

$$D_\ell C_0 = N_0 D_\ell \left(\frac{C_0}{N_0} \right) + C_0 D_\ell \ln N_0 = N_0 D_\ell \mathbf{E} \, \text{payoff}\, C + C_0 D_\ell \ln N_0,$$

$$\textit{with} \quad D_\ell \ln N_0 = 0 \quad \textit{for } \mathbb{P}_0 \quad \textit{and} \quad D_\ell \ln N_0 = -\frac{\delta_\ell}{1+\delta_\ell K(0,T_\ell)} \quad \textit{for } \mathbb{P}_n.$$

Knowing all the *option deltas* $D_\ell C_0$, or equivalently the *discounted option deltas* $D_\ell \left(\frac{C_0}{N_0}\right)$, permits sensitivities with respect to any other yieldcurve dependent instrument like swaps to found with a little extra work (see Remark-11.1).

One reasonably accurate method of computing $D_\ell C_0$ is to *bump and grind*: price the option, bump the forward, reprice and difference, that is, set

$$\begin{aligned} D_\ell \left(\frac{C_0}{N_0} \right) &\cong \frac{1}{\varepsilon} \left\{ \mathbf{E} \, \text{payoff}\, C \left(K(0,T_\ell) + \varepsilon \right) - \mathbf{E} \, \text{payoff}\, C \left(K(0,T_\ell) \right) \right\}, \\ &= \frac{1}{\varepsilon} \mathbf{E} \left\{ \text{payoff}\, C \left(K(0,T_\ell) + \varepsilon \right) - \text{payoff}\, C \left(K(0,T_\ell) \right) \right\}, \\ &\to \quad \mathbf{E} D_\ell \, \text{payoff}\, C \left(K(0,T_\ell) \right) \quad \textit{as} \quad \varepsilon \to 0, \end{aligned}$$

which allows the delta to be computed as the average of trajectory by trajectory differences. More accurate is the *pathwise method*: differentiate through the expectation and analytically compute $D_\ell \, \text{payoff}\, C(K(0,T_\ell))$ in

$$D_\ell \left(\frac{C_0}{N_0} \right) = D_\ell \mathbf{E} \, \text{payoff}\, C \left(.., K(0,T_\ell), .. \right) = \mathbf{E} D_\ell \, \text{payoff}\, C \left(.., K(0,T_\ell), .. \right).$$

That is, compute the partial derivative of the payoff and then average that derivative over all trajectories. We now develop a suit of methods for finding D_ℓ payoff $C(\cdot)$ for forward, zero, swap and option payoffs.

11.1 Partial derivatives of forwards

Because the shift is a function of maturity only, for $j, \ell = 0, .., n-1$, clearly

$$\frac{\partial}{\partial K(0,T_\ell)} = \frac{\partial}{\partial H(0,T_\ell)} \quad \textit{and}$$
$$\Delta_{j,\ell}(t) = D_\ell K(t,T_j) = \frac{\partial K(t,T_j)}{\partial K(0,T_\ell)} = \frac{\partial H(t,T_j)}{\partial H(0,T_\ell)}.$$

Recall from (5.1) and (5.7) that the SDE for $H(t,T_j)$ has form

$$\frac{dH(t,T_j)}{H(t,T_j)} = \mu_j(t)\,dt + \xi^*(t,T_j)\,dW(t) \qquad \Rightarrow \tag{11.1}$$
$$d\Delta_{j,\ell}(t) = \Delta_{j,\ell}(t)\frac{dH(t,T_j)}{H(t,T_j)} + H(t,T_j)\sum_{j1=1}^{n}\frac{\partial\mu_j(t)}{\partial K(0,T_{j1})}\Delta_{j1,\ell}(t)\,dt,$$
$$\textit{with initial condition} \qquad \Delta_{i,\ell}(0) = \mathbb{I}\{i=\ell\}.$$

Solving (11.1) numerically along with the Glasserman SDEs of Section-9.1 is too numerically intensive, and ignoring the drift leading to the solution

$$\Delta_{j,\ell}(t) = \frac{K(t,T_j)}{K(0,T_\ell)}\mathbb{I}[j=\ell],$$

is too biased. So use the low variance property of $h_j(t)$, see (3.4), and set

$$h_j(t) = h_j(0) = \frac{\delta_j H(0,T_j)}{[1+\delta_j K(0,T_j)]}$$

leading to the following approximate solution for $H(t,T_j)$ under $\mathbb{P}_0$

$$\frac{H(t,T_j)}{H(0,T_j)} = \mathcal{E}\left\{\int_0^t \mu_j^{(0)}(t)\,dt + \int_0^t \xi^*(t,T_j)\,dW_0(t)\right\}, \tag{11.2}$$
$$\textit{where} \qquad \mu_j^{(0)}(t) = \xi^*(t,T_j)\sum_{j1=@}^{j} h_{j1}(0)\,\xi(t,T_{j1}),$$

and an alternative approximation under the terminal measure $\mathbb{P}_n$

$$\frac{H(t,T_j)}{H(0,T_j)} = \mathcal{E}\left\{\int_0^t \mu_j^{(n)}(t)\,dt + \int_0^t \xi^*(t,T_j)\,dW_n(t)\right\}, \tag{11.3}$$

$$\textit{where} \qquad \mu_j^{(n)}(t) = -\xi^*(t,T_j)\sum_{j1=j+1}^{n-1} h_{j1}(0)\,\xi(t,T_{j1}).$$

Partially differentiating $H(t,T_j)$ with respect to $H(0,T_\ell)$, or equivalently $K(0,T_\ell)$, then yields

$$\Delta_{j,\ell}(t) = \frac{H(t,T_j)}{H(0,T_\ell)}\mathbb{I}\{j=\ell\} + H(t,T_j)\int_0^t D_\ell\mu_j^{(\cdot)}(s)\,ds, \tag{11.4}$$

$$D_\ell\mu_j^{(0)}(t) = \xi(t,T_j)\,\xi^*(t,T_\ell)\,D_\ell h_\ell(0)\,\mathbb{I}[@(t)\le\ell\le j],$$

$$D_\ell\mu_j^{(n)}(t) = -\xi(t,T_j)\,\xi^*(t,T_\ell)\,D_\ell h_\ell(0)\,\mathbb{I}[j<\ell<n], \quad \textit{with}$$

$$D_\ell h_\ell(0) = \frac{1-\delta_\ell a(T_\ell)}{[1+\delta_\ell K(0,T_\ell)]^2}.$$

11.2 Partial derivatives of zeros and swaps

From (9.1) the discounted swap S_{j0} at node $t=T_{j0}$ is

$$S(T_{j0}) = \frac{\text{pSwap}(T_{j0})}{N(T_{j0})}$$

$$= Z(T_{j0},T_{j0}) - Z(T_{j0},T_{jN}) + \sum_{j=j0}^{jN-1}\delta_j\mu_j Z(T_{j0},T_{j+1}) - \kappa\sum_{i=i0}^{iM-1}\overline{\delta}_i Z\left(T_{j0},\overline{T}_{i+1}\right),$$

so if we can find $D_\ell Z(T_{j0},T_k)$ $(k\ge j0)$, then we have $D_\ell S(T_{j0})$. Under $\mathbb{P}_0$

$$Z(T_{j0},T_k) = \frac{B(T_{j0},T_k)}{M(T_{j0})} = \left\{\begin{array}{c}\prod_{j=0}^{j0-1}\frac{1}{[1+\delta_j K(T_j,T_j)]}\\ \times\prod_{j=j0}^{k-1}\frac{1}{[1+\delta_j K(T_{j0},T_j)]}\end{array}\right\} \Rightarrow$$

$$D_\ell Z(T_{j0},T_k) = -Z(T_{j0},T_k)\left\{\begin{array}{c}\sum_{j=0}^{j0-1}\frac{\delta_j D_\ell K(T_j,T_j)}{[1+\delta_j K(T_j,T_j)]}\\ +\sum_{j=j0}^{k-1}\frac{\delta_j D_\ell K(T_{j0},T_j)}{[1+\delta_j K(T_{j0},T_j)]}\end{array}\right\},$$

while under $\mathbb{P}_n$

$$Z(T_{j0}, T_k) = \frac{B(T_{j0}, T_k)}{B(T_{j0}, T_n)} = \prod_{j=k}^{n-1} [1 + \delta_j K(T_{j0}, T_j)] \Rightarrow$$

$$D_\ell Z(T_{j0}, T_k) = Z(T_{j0}, T_k) \sum_{j=k}^{n-1} \frac{\delta_j D_\ell K(T_{j0}, T_j)}{[1 + \delta_j K(T_{j0}, T_j)]}.$$

REMARK 11.1 Letting the present values of the first N swaps be

$$\text{pSwap}(j) = 1 - B(0, T_j) - \kappa \sum_{\ell=0}^{j-1} \delta_\ell B(0, T_{\ell+1}), \qquad j = 1, .., N,$$

the Jacobian $J = (J_{j,\ell}) = \left(\frac{\partial \, \text{pSwap}(j)}{\partial K(0,T_l)} \right)$ can be found and inverted, allowing the *swap deltas* $\frac{\partial C_0}{\partial \, \text{pSwap}(j)}$ to be expressed in terms of the *forward deltas.* □

11.3 Differentiating option payoffs

Following are some not very rigorously derived results that are useful in differentiating option payoffs, see Section-A.5 for further information.

Option Greeks usually involve the *Heaviside* (or characteristic) function $\mathbb{I}(\cdot)$ and the *positive value function* $(\cdot)^+$ defined respectively by

$$\mathbb{I}(x) = \begin{cases} 1 & \text{when} \quad x > 0 \\ \frac{1}{2} & \text{when} \quad x = 0 \\ 0 & \text{when} \quad x < 0 \end{cases}, \qquad (x)^+ = \max(0, x) = x \, \mathbb{I}(x),$$

and also the *Dirac delta function* $\boldsymbol{\delta}(\cdot)$, a probability measure such that

$$\int_a^b \boldsymbol{\delta}(y) \, dy = \begin{cases} 1 & \text{when} \quad 0 \in (a, b) \\ 0 & \text{when} \quad 0 \notin (a, b) \end{cases}.$$

Integrating these functions yields expressions for their derivatives

$$\int_{-\infty}^x \mathbb{I}(u) \, du = (x)^+ \qquad \Rightarrow \qquad \frac{d}{dx}(x)^+ = \mathbb{I}(x), \tag{11.5}$$

$$\int_{-\infty}^x \boldsymbol{\delta}(u) \, du = \mathbb{I}(x) \qquad \Rightarrow \qquad \frac{d}{dx}\mathbb{I}(x) = \boldsymbol{\delta}(x),$$

moreover, writing $(x)^+ = x \, \mathbb{I}(x)$ and differentiating gives

$$\begin{aligned} \frac{d}{dx}(x)^+ &= \frac{d}{dx} x \, \mathbb{I}(x) = \mathbb{I}(x) + x \, \boldsymbol{\delta}(x) = \mathbb{I}(x) \\ \Rightarrow & \qquad x \boldsymbol{\delta}(x) = 0. \end{aligned} \tag{11.6}$$

REMARK 11.2 $\mathbb{I}(0) = \frac{1}{2}$ is consistent with $\boldsymbol{\delta}(\cdot)$ being an even function, and follows Fourier transform practice, see Section-A.5. □

The next lemma, which follows from (11.6), deals with a recurring kind of expression:

LEMMA 11.1

If $X = X^\varepsilon$, $Y = Y^\varepsilon$, $Z = Z^\varepsilon$ are random variables dependent on a parameter ε, and $\mathcal{A} = \mathcal{A}\{X \geq Y\}$ is the event that $X \geq Y$, then

$$\mathbf{E}\left\{Z\left[X\,\frac{\partial}{\partial\varepsilon}\mathbb{I}(\mathcal{A}) + Y\,\frac{\partial}{\partial\varepsilon}\mathbb{I}(\mathcal{A}^c)\right]\right\} = \mathbf{E}\{Z(X-Y)\;\boldsymbol{\delta}(X-Y)\} = 0.$$

REMARK 11.3 In practice, when the characteristic function represents the payoff of a digital option, it is usually approximated by a *call straddle*

$$\mathbb{I}(x) \cong \frac{1}{\varepsilon}\left\{(x)^+ - (x-\varepsilon)^+\right\} \quad \Rightarrow \quad \boldsymbol{\delta}(x) \cong \frac{1}{\varepsilon}\mathbb{I}[0 \leq x \leq \varepsilon],$$

which expression is useful for simulating the Dirac delta function according to

$$\mathbf{E}\;Z\;\boldsymbol{\delta}(X) = \frac{1}{\varepsilon}\;\mathbf{E}\;Z\;\mathbb{I}[0 \leq X \leq \varepsilon]$$

for random variables X and Z. □

11.4 Vanilla caplets and swaptions

The deltas of vanilla caplets and swaptions can of course be found by analytic methods, and in the case of a caplet, exactly. That provides a way of checking, see [44], the accuracy of the pathwise delta method.

For a T-maturing caplet, the discounted payoff at T_1 is

$$\text{payoff}\,C = \frac{\delta\,(K(T,T)-\kappa)^+}{N(T_1)} = [Z(T,T) - (1+\delta\kappa)\,Z(T,T_1)]^+ \quad \Rightarrow$$

$$D_\ell\,\text{payoff}\,C = \left\{\begin{array}{c} \mathbb{I}[Z(T,T) - (1+\delta\kappa)\,Z(T,T_1)] \\ \times\,[D_\ell Z(T,T) - (1+\delta\kappa)\,D_\ell Z(T,T_1)] \end{array}\right\},$$

while for a T_{j0}-maturing swaption, the discounted payoff at T_{j0} is

$$\text{payoff } C = [S(T_{j0})]^+ = \begin{bmatrix} Z(T_{j0}, T_{j0}) - Z(T_{jN}) \\ -\kappa \sum_{i=i0}^{iM-1} \overline{\delta}_i Z(T_{j0}, \overline{T}_{i+1}) \end{bmatrix}^+ \Rightarrow$$

$$D_\ell \text{ payoff } C = \mathbb{I}[S(T_{j0})] \begin{bmatrix} D_\ell Z(T_{j0}, T_{j0}) - D_\ell Z(T_{j0}, T_{jN}) \\ -\kappa \sum_{i=i0}^{iM-1} \overline{\delta}_i D_\ell Z(T_{j0}, \overline{T}_{i+1}) \end{bmatrix}.$$

Finish by substituting for the $D_\ell Z(\cdot)$ and $D_\ell K(\cdot)$.

11.5 Barrier caps and floors

A *barrier cap or floor* is a vanilla cap knocked out or in by the current forward crossing a barrier. Because a *knock-in* together with a *knock-out* is the same as the underlying option, barrier options will be generally cheaper than their vanilla equivalents. Hence their attraction to customers who are willing to *take a view* in exchange for lower premiums. An illustration is the following:

Example 11.1
With Libor at 5%, a customer takes a loan paying floating and for protection buys for about \$10k a cap struck at 7.4%. But he doesn't think rates will go above 6% and in any case can tolerate 7.4%, so he sells for about \$100k an up-and-in floor struck at 7.4% with barrier at 6%. If his view is right, he has reduced the cost of his loan by \$90k, if wrong he'll pay 7.4% when floating is between 6% and 7.4%. □

As with all other barrier options, for example, FX barriers, there are sixteen flavours corresponding to combinations of cap or floor, up or down, in or out, and barrier less or greater than strike.

Value a barrier floor struck at κ with barrier β ($< \kappa$), by valuing the component floorlets C_k ($k = 1, .., N$) fixed at T_k and paying at T_{k+1}. Let

$$\mathcal{A}_j = \{K(T_j, T_j) \geq \beta\}$$

be the event (complement $\mathcal{A}_j^c$) that the barrier is crossed at T_j. Then the floorlet payoff is

$$\text{payoff } C_k = \mathbf{E}\left\{ \begin{array}{c} \mathbb{I}[\mathcal{A}_1] + \mathbb{I}[\mathcal{A}_1^c]\mathbb{I}[\mathcal{A}_2] + \mathbb{I}[\mathcal{A}_1^c]\mathbb{I}[\mathcal{A}_2^c]\mathbb{I}[\mathcal{A}_3] \\ +... + \mathbb{I}[\mathcal{A}_1^c]\mathbb{I}[\mathcal{A}_2^c]\mathbb{I}[\mathcal{A}_3^c]..\mathbb{I}[\mathcal{A}_{k-1}^c]\mathbb{I}[\mathcal{A}_k] \end{array} \right\} \frac{\delta(\kappa - K(T_k, T_k))^+}{N(T_{k+1})},$$

$$= \mathbf{E}\left\{ \begin{array}{c} \mathbb{I}[\mathcal{A}_1] + \mathbb{I}[\mathcal{A}_1^c]\mathbb{I}[\mathcal{A}_2] + ... \\ \mathbb{I}[\mathcal{A}_1^c]\mathbb{I}[\mathcal{A}_2^c]..\mathbb{I}[\mathcal{A}_{k-1}^c]\mathbb{I}[\mathcal{A}_k] \end{array} \right\} [(1 + \delta\kappa) Z(T_k, T_{k+1} - Z(T_k, T_k))]^+,$$

and its delta is a rather complex expression into which the $D_\ell Z(\cdot)$ and $D_\ell K(\cdot)$ from the previous sections must be substituted

$$D_\ell \,\text{payoff}\, C_k = \mathbf{E}\left\{\begin{array}{c} D_\ell K(T_1,T_1)\,\boldsymbol{\delta}[\mathcal{A}_1]\left(\begin{array}{c} 1-\mathbb{I}[\mathcal{A}_2]-\mathbb{I}[\mathcal{A}_2^c]\,\mathbb{I}[\mathcal{A}_3]-... \\ -\mathbb{I}[\mathcal{A}_2^c]\,\mathbb{I}[\mathcal{A}_3^c]\,..\mathbb{I}\left[\mathcal{A}_{k-1}^c\right]\mathbb{I}[\mathcal{A}_k]\end{array}\right) \\ +D_\ell K(T_2,T_2)\,\boldsymbol{\delta}[\mathcal{A}_2]\left(\begin{array}{c} \mathbb{I}[\mathcal{A}_1^c]-\mathbb{I}[\mathcal{A}_1^c]\,\mathbb{I}[\mathcal{A}_3]-... \\ -\mathbb{I}[\mathcal{A}_1^c]\,\mathbb{I}[\mathcal{A}_3^c]\,..\mathbb{I}\left[\mathcal{A}_{k-1}^c\right]\mathbb{I}[\mathcal{A}_k]\end{array}\right) \\ \vdots \end{array}\right\}$$

$$\times\left[(1+\delta\kappa)\,Z(T_k,T_{k+1}-Z(T_k,T_k))\right]^+$$

$$+\left\{\begin{array}{c}\mathbb{I}[\mathcal{A}_1]+\mathbb{I}[\mathcal{A}_1^c]\,\mathbb{I}[\mathcal{A}_2]+\mathbb{I}[\mathcal{A}_1^c]\,\mathbb{I}[\mathcal{A}_2^c]\,\mathbb{I}[\mathcal{A}_3] \\ +...+\mathbb{I}[\mathcal{A}_1^c]\,\mathbb{I}[\mathcal{A}_2^c]\,\mathbb{I}[\mathcal{A}_3^c]\,..\mathbb{I}\left[\mathcal{A}_{k-1}^c\right]\mathbb{I}[\mathcal{A}_k]\end{array}\right\}$$

$$\times\left\{\begin{array}{c}\mathbb{I}\left[(1+\delta\kappa)\,Z(T_k,T_{k+1}-Z(T_k,T_k))\right] \\ \times\left[D_\ell Z(T,T)-(1+\delta\kappa)\,D_\ell Z(T,T_1)\right]\end{array}\right\}.$$

In simulating this expression, the Dirac delta functions are of course handled along the lines of Remark-11.3.

Chapter 12

Bermudans

The owner of a *payer (receiver) Bermudan* swaption has the right to exercise into an underlying payer (receiver) swap at some subset of its fixed side reset dates.

A good source of revenue for banks is an arrangement popular with investors, in which they receive higher than usual fixed interest in exchange for giving the bank the right to cancel the deal if it does not suit. At the centre of the structure is a *callable swap*, which is a payer swap that can be cancelled at any of its reset dates. A callable swap in which the bank pays fixed, is thus equivalent to the bank owning a vanilla payer swap plus a receiver Bermudan, because if the bank exercises the Bermudan, it gets a receiver swap that offsets (cancels) the remainder of the payer swap.

Roughly speaking, these sorts of deals are set up as follows. The investor deposits funds with the bank which are safely invested and generate floating Libor back to the bank. The bank then sets up (with its exotic derivatives desk) a callable swap in which the exotics desk receives that floating Libor and pays a fixed coupon. But because the swap is cancellable, that is equivalent to the exotics desk getting a free receiver Bermudan, which is worth money. That free money is then in part used to increase the fixed coupon in the underlying payer swap, which goes back to the investor as enhanced yield.

A moment's thought reveals that such deals will tend to get cancelled if Libor rates fall, because the payer swap is then getting more expensive for the exotics desk. So the investor *takes a view* and bets that interest rates will remain steady; if he is right, he receives enhanced coupon for the full duration of the deposit, if he is wrong, he gets his deposit back early and must find an alternative investment at probably lower rates.

Different *views* for the consideration of the investor can be created by substituting *exotic coupons* for the fixed coupon. For example, a *callable constant maturity spread swap* might, instead of the fixed coupon, pay a constant coupon plus a positive multiple of the 2-year constant maturity swaprate less the 30-year constant maturity swaprate. The investor receives enhanced yield betting that long-term rates don't fall relative to short-term rates (because in that case the exotics desk must pay increased coupon perhaps causing them to cancel).

Receiver Bermudans are also often used to partially hedge fixed coupon mortgaged backed securities (MBS); if rates drop and mortgagees refinance at a lower coupon, banks may still want to receive the original higher coupon.

The huge size of the US MBS market thus accounts for the massive Bermudan books carried by many US banks.

Bermudans are now so common as to be almost vanilla products; some traders calibrate to them! Nevertheless, because they are the *archetypical callable product*, their pricing exhibits many of the techniques needed to price more esoteric *callable exotics*; hence this chapter. They are also very stable instruments, relatively easy to price and hedge, and not overly sensitive to correlation or the swaption volatility smile.

We now show how to combine Glasserman style simulation with Longstaff-Schwartz's regression technique [73] for conditional expectations, to value a *payer Bermudan* which can be exercised into a notional underlying fixed maturity payer swap at any one of its $(M-1)$ reset times $\overline{T}_i$ $(i=0,1,...,M-1)$. The author would also suggest Chapter-8 of Glasserman's book [46] for parallel reading with this chapter.

12.1 Backward recursion

Denote by $\mathbb{P}$ (numeraire $N(t)$ expectation $\mathbf{E}$) whichever measure (spot $\mathbb{P}_0$ or terminal $\mathbb{P}_n$) we are using, let $S(t)$ be the time-t *discounted to* $t=0$ *intrinsic* value of the underlying swap with next reset $\overline{T}_i$

$$S(t) = \frac{\text{pSwap}\left(t, \overline{T}_i, \overline{T}_M\right)}{N(t)},$$

so that at a reset when $t = \overline{T}_i$ we can use the notation

$$S_i = S\left(\overline{T}_i\right) = \frac{\text{pSwap}\left(\overline{T}_i, \overline{T}_i, \overline{T}_M\right)}{N\left(\overline{T}_i\right)}, \tag{12.1}$$

and write expectations conditioned on $\mathcal{F}_i = \mathcal{F}_{\overline{T}_i}$ as

$$\mathbf{E}^{(i)} X = \mathbf{E}\left\{ X \middle| \mathcal{F}_i \right\}.$$

REMARK 12.1 Here, and generally throughout this chapter, we work with variables like the discounted $S_i = S\left(\overline{T}_i\right)$ in equation (12.1) rather than the undiscounted $\text{pSwap}\left(t, \overline{T}_i, \overline{T}_M\right)$. That is, values of all relevant instruments will be discounted by the numeraire $N(t)$ to time $t=0$ making them $\mathbb{P}$-martingales. Note also that any exercise decisions determined by inequalities can be based on either non-discounted or discounted values; because numeraires are positive the results must be the same. □

The present value $\text{Bm}(0)$ of a Bermudan is therefore the supremum taken over all *discrete* stopping times $\tau \in \overline{\mathcal{T}} = \left\{\overline{T}_i \ : i=0,1,...,M-1\right\}$ of potential discounted payoffs, namely

$$\mathrm{Bm}\,(0) = \sup_{\tau \in \overline{\mathcal{T}}} \mathbf{E} S\,(\tau) = \mathbf{E} S\,(\tau^*)\,, \tag{12.2}$$

where $\tau^* \in \overline{\mathcal{T}}$ is the optimal stopping time.

If $\mathrm{Bm}\,(t)$ is the *discounted to* $t = 0$ *Bermudan* value at time t, then at consecutive resets $\overline{T}_i$ clearly $\mathrm{Bm}\left(\overline{T}_i\right)$ is the maximum of the *intrinsic value* $S\left(\overline{T}_i\right)$ and the *continuation value* $\mathbf{E}\left\{\mathrm{Bm}\left(\overline{T}_{i+1}\right)\middle|\,\mathcal{F}_{\overline{T}_i}\right\}$

$$\mathrm{Bm}\left(\overline{T}_i\right) = \max\left[S\left(\overline{T}_i\right),\ \mathbf{E}\left\{\mathrm{Bm}\left(\overline{T}_{i+1}\right)\middle|\,\mathcal{F}_{\overline{T}_i}\right\}\right],$$

which, using the above notation and setting $\mathrm{Bm}\left(\overline{T}_i\right) = \mathrm{Bm}_i$, can be written more compactly as

$$\mathrm{Bm}_i = S_i \vee \mathbf{E}^{(i)}\,\mathrm{Bm}_{i+1}\,. \tag{12.3}$$

Hence $\mathrm{Bm}\,(t)$ is a supermartingale that dominates $S\,(t)$ because both

$$\mathrm{Bm}\left(\overline{T}_i\right) \geq \mathbf{E}\left\{\mathrm{Bm}\left(\overline{T}_{i+1}\right)\middle|\,\mathcal{F}_{\overline{T}_i}\right\} \quad \textit{and} \quad \mathrm{Bm}\left(\overline{T}_i\right) \geq S\left(\overline{T}_i\right).$$

Repeated application of (12.3) for $i = M-1,..,0$ then generates the following *backward recursion* for $\mathrm{Bm}\,(0) = \mathrm{Bm}_0$

$$\begin{aligned} \mathrm{Bm}_{M-1} &= S_{M-1} \vee 0 = S_{M-1}^{+}, \\ \mathrm{Bm}_i &= S_i \vee \mathbf{E}^{(i)}\,\mathrm{Bm}_{i+1}, \quad (i = M-2,..,0)\,. \end{aligned} \tag{12.4}$$

Letting $\mathcal{A}_i$ be the event that at $\overline{T}_i$ the *intrinsic value* $S_i = S\left(\overline{T}_i\right)$ of the underlying swap is greater than the *continuation value* $\mathbf{E}^{(i)}\,\mathrm{Bm}_{i+1}$, that is

$$\begin{aligned} \mathcal{A}_i &= \left\{S_i \geq \mathbf{E}^{(i)}\,\mathrm{Bm}_{i+1}\right\}, \quad i = 0,1,..,M-2 \\ &\textit{and} \qquad \mathcal{A}_{M-1} = \{S_{M-1} \geq 0\}\,, \end{aligned}$$

the recursion (12.4) combines to give

$$\begin{aligned} \mathrm{Bm}_0 &= S_0 \vee \mathbf{E}^{(0)}\left[S_1 \vee \mathbf{E}^{(1)}\left[S_2 \vee \mathbf{E}^{(2)}\left[...\left[S_{M-2} \vee \mathbf{E}^{(M-2)}\left[S_{M-1} \vee 0\right]\right]\right]\right]\right] \\ &= \mathbf{E}^{(0)}\mathbf{E}^{(1)}\mathbf{E}^{(2)}...\mathbf{E}^{(M-2)} S_0 \vee S_1 \vee S_2 \vee S_3 \vee S_4 \vee ....S_{M-2} \vee S_{M-1} \vee 0, \\ &= \mathbf{E}\ S_0 \vee S_1 \vee S_2 \vee S_3 \vee ....S_{M-2} \vee S_{M-1}^{+}, \\ &= \mathbf{E}\left\{\begin{array}{c} \mathbb{I}\left[\mathcal{A}_0\right] S_0 + \mathbb{I}\left[\mathcal{A}_0^c\right]\mathbb{I}\left[\mathcal{A}_1\right] S_1 + \mathbb{I}\left[\mathcal{A}_0^c\right]\mathbb{I}\left[\mathcal{A}_1^c\right]\mathbb{I}\left[\mathcal{A}_2\right] S_2 + .. \\ ... + \mathbb{I}\left[\mathcal{A}_0^c\right]\mathbb{I}\left[\mathcal{A}_1^c\right]..\mathbb{I}\left[\mathcal{A}_{M-2}^c\right]\mathbb{I}\left[\mathcal{A}_{M-1}\right] S_{M-1} \end{array}\right\}. \end{aligned}$$

Now our problem is to decide which of S_i or $\mathbf{E}^{(i)}\,\mathrm{Bm}_{i+1}$ is greater at reset time $\overline{T}_i$, in the context of a simulation algorithm with a large number of simulated trajectories, and that is where the Longstaff-Schwartz technique is appropriate. For simplicity, we will illustrate his method via an example in Section-12.2 below.

12.1.1 Alternative backward recursion

For a more numerically efficient backward recursion introduce

$$H_i = H\left(\overline{T}_i\right) = \mathbf{E}^{(i)}\,\mathrm{Bm}_{i+1} - S_i, \qquad H_{M-1} = -S_{M-1}.$$

Using the recurrence (12.4) for Bm_i, the recurrence between H_{i-1} and H_i for $i = M-1,..,1$ is therefore

$$\begin{aligned} H_{i-1} &= \mathbf{E}^{(i-1)}\left\{S_i \vee [H_i + S_i]\right\} - S_{i-1} = \mathbf{E}^{(i-1)}\left\{S_i + (H_i)^+\right\} - S_{i-1}, \\ &= \mathbf{E}^{(i-1)}\,(H_i)^+ + \mathbf{E}^{(i-1)}\left\{S_i\right\} - S_{i-1}. \end{aligned}$$

But from (9.1) and (12.1)

$$\begin{aligned} S_i &= 1 - Z\left(\overline{T}_i, \overline{T}_M\right) - \kappa \sum_{i1=i}^{M-1} \overline{\delta}_{i1} Z\left(\overline{T}_i, \overline{T}_{i1+1}\right), \\ \Rightarrow \qquad & \mathbf{E}^{(i-1)}\left\{S_i\right\} - S_{i-1} = \kappa \overline{\delta}_{i-1} Z\left(\overline{T}_{i-1}, \overline{T}_i\right). \end{aligned}$$

Hence a recursion involving only one coupon

$$\begin{aligned} H_{M-1} &= -S_{M-1}, \qquad B_0 = H_0 + S_0, \\ H_{i-1} &= \mathbf{E}^{(i-1)}\left\{(H_i)^+\right\} + \kappa \overline{\delta}_{i-1} Z\left(\overline{T}_{i-1}, \overline{T}_i\right) \quad i = M-1,..,1 \end{aligned}$$

as opposed to the many time consuming coupon calculations needed in (12.4).

REMARK 12.2 If the coupon payment κ in the underlying swap $S(t)$ is exotic it may be difficult to compute the intrinsic values $S_i = S\left(\overline{T}_i\right)$, a problem that this algorithm solves because to compute H_{i-1} only the immediate coupon determined at $\overline{T}_{i-1}$ is required. An alternative method of computing, or rather estimating, intrinsic values $S_i = S\left(\overline{T}_i\right)$ for exotic coupon swaps is to apply the Longstaff-Schwartz regression technique of the next Section-12.2 and regress realized coupon values along each trajectory against values of relevant variables at time $\overline{T}_i$. □

12.2 The Longstaff-Schwartz lower bound technique

We illustrate Longstaff-Schwartz's method of computing lower bound values of callable options like Bermudans, through an example. Consider a 10-year Bermudan struck at κ which can be exercised annually into a 10-year swap with annual rolls; that is, $M = 10$ and $\overline{T}_i = i$.

12.2.1 When to exercise

On each of $\mathcal{K}$ simulated trajectories ω_k $(k = 1, .., \mathcal{K})$, the Bermudan could be exercised at just one of the 9 reset times $\overline{T}_1, .., \overline{T}_9$, say $\overline{T}_i$, into one of the following 9 swaps with discounted *intrinsic* value

$$I\left(\omega_k, \overline{T}_i\right) = S_i = S\left(\overline{T}_i\right) = \frac{\text{pSwap}\left(\overline{T}_i, \overline{T}_i, \overline{T}_{10}\right)}{N\left(\overline{T}_i\right)} \qquad i = 1, .., 9.$$

For trajectory ω_k suppose $\tau(\omega_k) \in \left\{\overline{T}_1, .., \overline{T}_9\right\}$ is the exercise time. Each value $I(\omega_k, \tau(\omega_k))$ is discounted to the root, so their average

$$\frac{1}{\mathcal{K}} \sum_k I(\omega_k, \tau(\omega_k)) \leq \text{Bm}(0)$$

will be a lower bound estimate of the Bermudan's price.

Proceed as follows to carry back along ω_k the up-to $\overline{T}_i$ *best discounted intrinsic value* $Y_k\left(\overline{T}_i\right)$ so far.

Step-1 Start with reset $\overline{T}_9 = 9$ on trajectory ω_k. Because exercise is determined solely by the intrinsic value $I\left(\omega_k, \overline{T}_9\right)$ at reset $\overline{T}_9$, clearly the corresponding best discounted value at $\overline{T}_9$ is

$$Y_k\left(\overline{T}_9\right) = Y_k\left(\omega_k, \overline{T}_9\right) = \begin{cases} I\left(\omega_k, \overline{T}_9\right) & \textit{if} \quad I\left(\omega_k, \overline{T}_i\right) > 0 \\ 0 & \textit{if} \quad I\left(\omega_k, \overline{T}_i\right) \leq 0 \end{cases}.$$

Step-2 Move to reset $\overline{T}_8$. On trajectory ω_k the choice for best discounted value $Y_k\left(\overline{T}_8\right)$ at $\overline{T}_8$ is either $I\left(\omega_k, \overline{T}_8\right)$ if it is best to exercise at $\overline{T}_8$ or $Y_k\left(\overline{T}_9\right)$ if it is best to exercise later. In general to step from $\overline{T}_{i+1}$ to $\overline{T}_i$ use the *continuation value* $\mathbf{E}\left(\text{Bm}\left(\overline{T}_{i+1}\right) \middle| \mathcal{F}_{\overline{T}_i}\right)$ to make the exercise decision, so that the best discounted value at $\overline{T}_i$ is

$$Y_k\left(\overline{T}_i\right) = \begin{cases} Y_k\left(\overline{T}_{i+1}\right) & \textit{if} \quad \begin{cases} I\left(\omega_k, \overline{T}_i\right) \leq 0 \quad \textit{or} \\ \left[\begin{array}{c} I\left(\omega_k, \overline{T}_i\right) > 0 \qquad \textit{and} \\ \mathbf{E}\left(\text{Bm}\left(\overline{T}_{i+1}\right) \middle| \mathcal{F}_{\overline{T}_i}\right)(\omega_k) > I\left(\omega_k, \overline{T}_i\right) \end{array}\right] \end{cases}, \\ I\left(\omega_k, \overline{T}_i\right) & \textit{if} \quad \left[\begin{array}{c} I\left(\omega_k, \overline{T}_i\right) > 0 \qquad \textit{and} \\ \mathbf{E}\left(\text{Bm}\left(\overline{T}_{i+1}\right) \middle| \mathcal{F}_{\overline{T}_i}\right)(\omega_k) \leq I\left(\omega_k, \overline{T}_i\right) \end{array}\right]. \end{cases}$$

Step-3 Repeat Step-2 through the resets $\overline{T}_7, \overline{T}_6, ..., \overline{T}_1$. On reaching the first reset $\overline{T}_1$ we have found a lower bound on $\text{Bm}(0)$

$$I(\omega_k, \tau(\omega_k)) = Y_k\left(\overline{T}_1\right) \qquad \Rightarrow \qquad \frac{1}{\mathcal{K}} \sum_k Y_k\left(\overline{T}_1\right) \leq \text{Bm}(0).$$

It remains to identify or approximate the continuation values.

12.2.2 Regression technique

Because our example is Markov in the 40 (quarterly for 10 years is 4×10) Glasserman $V(\cdot)$ variables, the information set $\mathcal{F}_{\overline{T}_i}$ will be determined by their time $\overline{T}_i$ values or any linearly independent combination thereof. So the conditional expectation that is the *continuation value* will be a function of those variables evaluated at $\overline{T}_i$.

$$\mathbf{E}\left(B\left(\overline{T}_{i+1}\right)\middle|\mathcal{F}_{\overline{T}_i}\right) = f_i\left(\overline{T}_i\right) = f_i\left(V_1\left(\overline{T}_i\right), V_2\left(\overline{T}_i\right), .., V_{40}\left(\overline{T}_i\right)\right)$$

and if we find or approximate $f_i\left(\overline{T}_i\right)$, we have the continuation values.

Using linear $\mathbb{L}^2$ regression, we estimate on timeslice $\overline{T}_i$ the corresponding continuation function $f_i\left(\overline{T}_i\right)$ from the information embodied in the simulated trajectories. Some robust heuristic approximation will be needed, because it is not practical to regress on a large number of basis functions. The art is to choose for basis functions those linear combinations of the Glasserman $V(\cdot)$ variables that account for most of the value of the Bermudan.

Reasonable results can be obtained by regressing on just the intrinsic values. For example, in stepping down from reset $\overline{T}_9 \to \overline{T}_8$ let

$$U = U(\omega_k) = \left[I\left(\omega_k, \overline{T}_8\right)\right]^+,$$

which is the positive part of the intrinsic value at time $\overline{T}_8$ and regress on $1, U$ and U^2. The constant 1 is included to add useful stability. Including a in the regression

$$\min_{a,b,c..} \sum_k \left(Y_k - a - bU_k - cU_k^2\right)^2,$$

makes the means of target and estimate equal on minimizing with respect to a

$$\frac{\partial}{\partial a} = 0 \quad \Rightarrow \quad \sum_k Y_k = \sum_k a + bU_k + cU_k^2.$$

The regression runs as follows:

Step-1 For negative intrinsic values $I(\omega_k, R_8)$ at reset $\overline{T}_8$ no decision is required, so first identify the set containing $\mathcal{K}(8)$ trajectories

$$\Omega_8^+ = \left\{\omega_k \ \ (k = 1, .., \mathcal{K}(8)) : I\left(\omega_k, \overline{T}_8\right) > 0\right\}$$

for which that intrinsic value $I\left(\omega_k, \overline{T}_8\right)$ is strictly positive. We will regress on the corresponding $U_k = U(\omega_k)$, where the $\omega_k \in \Omega_8^+$ for $k = 1, 2, .., \mathcal{K}(8)$.

Step-2 Look for the best solution $C = \left(c_1\ c_2\ c_3\right)^T$ in the $\mathbb{L}^2$ sense to the over-specified system of equations

$$\begin{pmatrix} 1 & U_1 & U_1^2 \\ \vdots & \vdots & \vdots \\ 1 & U_k & U_k^2 \\ \vdots & \vdots & \vdots \\ 1 & U_{\mathcal{K}(8)} & U_{\mathcal{K}(8)}^2 \end{pmatrix} C = \begin{pmatrix} Y_1\left(\overline{T}_9\right) \\ \vdots \\ Y_k\left(\overline{T}_9\right) \\ \vdots \\ Y_{\mathcal{K}(8)}\left(\overline{T}_9\right) \end{pmatrix}.$$

Step-3 Having found C, compute the estimates

$$f_8\left(\overline{T}_8\right) = \mathbf{E}\left(B\left(\overline{T}_9\right)\middle|\mathcal{F}_8\right)$$
$$= \begin{pmatrix} \mathbf{E}\left(B\left(\overline{T}_9\right)\middle|\mathcal{F}_8\right)(\omega_1) \\ \vdots \\ \mathbf{E}\left(B\left(\overline{T}_9\right)\middle|\mathcal{F}_8\right)(\omega_k) \\ \vdots \\ \mathbf{E}\left(B\left(\overline{T}_9\right)\middle|\mathcal{F}_8\right)\left(\omega_{\mathcal{K}(8)}\right) \end{pmatrix} \cong \begin{pmatrix} 1 & U_1 & U_1^2 \\ \vdots & \vdots & \vdots \\ 1 & U_k & U_k^2 \\ \vdots & \vdots & \vdots \\ 1 & U_{\mathcal{K}(8)} & U_{\mathcal{K}(8)}^2 \end{pmatrix} C$$

of the conditional expectations $\mathbf{E}\left(B\left(\overline{T}_9\right)\middle|\mathcal{F}_8\right)(\omega_k)$ for each of the trajectories $\omega_k \in \Omega_8^+$ with positive intrinsic values $I\left(\omega_k, \overline{T}_8\right) > 0$.

Step-4 Repeat Step-3 backwards through the resets $\overline{T}_7, \overline{T}_6, ..., \overline{T}_1$.

A good technique to reduce bias (see [46] for further comment), is to use one set of trajectories (about 1k is enough) to compute the regression coefficients (and thus effectively the stopping times), and another set (4k - 10k) to compute prices.

12.2.3 Comments on the Longstaff-Schwartz technique

The Longstaff-Schwartz method clearly gives a lower bound on the price of a payer Bermudan, because optimizing with a restricted set of regression variables is an approximation that might be improved with a larger and better choice. The author's general experience with the technique, however, is that it is robust and with proper choice of regression variables the lower bound obtained is an accurate measure of the Bermudans value.

The choice of regression variables is partly an art and partly a science, and generally they should be tailored to the callable instrument being valued and involve variables that in a linear combination can mimic its potential behavior.

Thus for vanilla Bermudans useful regression variables might include: swap values, because they are linear combinations of the finite number of Markov variables (the Glasserman $V(\cdot)$ variables) underlying the model, and so make good variables for regression; zero coupons to reflect the level of the yieldcurve; differences of zero coupons to reflect tilt and flex in the yieldcurve; European swaptions maturing at the reset dates, because they form lower bounds etc.

The author has tended to rely on a combination of judicious use of the timeslicer (recall that it does not incorporate correlation but is otherwise accurate) to check simulation results, and experiment with different sets of regression variables (trying to push up prices and get them stable) to increase his comfort levels with the lower bound Longstaff-Schwartz prices. He has yet to implement the time-consuming upper bound method described in Section-12.3 below, and thus get a genuine confidence interval for prices. Moreover, at the present moment he can't help feeling that perhaps the proper place for the upper bound method is more as a risk management tool to check models, rather than an everyday front office pricer.

12.3 Upper bounds

The lower bound on the Bermudan price as obtained by Longstaff-Schwartz's technique corresponds to the buyer's bid. He owns the option, has perhaps paid less than the optimal price for it, and can decide to exercise it according to the optimal routine he has used to compute his bid, or indeed at any other time of his choosing.

An upper bound on the Bermudan price corresponds naturally to the seller's offer, he must hedge the sold Bermudan and cater for the possibility that the buyer might exercise at any time, including perhaps accidentally at a time that is more advantageous to him than the optimal time.

A series of papers have tackled the upper bound problem starting with essentially equivalent independent approaches by Rogers [109] and Haugh and Kogan [51], and an alternative approach by Jamshidian [67]. Subsequently Andersen and Broadie [8] added operational depth, which was refined by Joshi [69], [70], [71].

From (12.2), for any stopping time $\tau \in \overline{\mathcal{T}}$

$$\mathbf{E}S(\tau) \leq \mathrm{Bm}(0) = \sup_{\tau\in\overline{\mathcal{T}}} \mathbf{E}S(\tau) = \mathbf{E}S(\tau^*),$$

showing that any stopping time can be used to compute a lower bound on the Bermudan price Bm_0.

To get an upper bound, let H be the set of all adapted martingales $h(t)$ for which $\sup_{t\in\overline{\mathcal{T}}} |h(t)| < \infty$. Then for any $h \in H$, we have

$$\begin{aligned}\mathrm{Bm}(0) &= \sup_{\tau\in\overline{\mathcal{T}}} \mathbf{E}[S(\tau) + h(t) - h(t)] = \mathbf{E}h(\tau^*) + \sup_{\tau\in\overline{\mathcal{T}}} \mathbf{E}[S(\tau) - h(t)],\\ &= h(0) + \sup_{\tau\in\overline{\mathcal{T}}} \mathbf{E}[S(\tau) - h(t)],\end{aligned}$$

from the optional sampling theorem applied to the martingale $h(t)$. Hence

$$\mathrm{Bm}(0) \leq h(0) + \mathbf{E}\max_{t\in\overline{\mathcal{T}}} [S(t) - h(t)], \tag{12.5}$$

where the maximum is now taken over the exercise times in

$$\overline{\mathcal{T}} = \left\{\overline{T}_i \ : i = 0, 1, ..., M-1\right\},$$

and not the stopping times. Because $h \in H$ was arbitrary, that yields a problem dual to (12.2), namely

$$\mathrm{Bm}(0) \leq \inf_{h\in H} \left\{h(0) + \mathbf{E}\max_{t\in\overline{\mathcal{T}}} [S(t) - h(t)]\right\}. \tag{12.6}$$

In fact, as Rogers showed, equality holds because $\mathrm{Bm}(t)$, being a supermartingale that dominates $S(t)$ according to the Doob-Meyer decomposition can be written

$$\mathrm{Bm}(t) = M(t) - A(t),$$

where $M(t)$ is a martingale and $A(t)$ is an increasing process with $A(0) = 0$. Substituting $M(t)$ for $h(t)$ in (12.5) gives

$$\mathrm{Bm}(0) \leq \mathrm{Bm}(0) + \mathbf{E}\max_{t\in\overline{T}}[S(t) - \mathrm{Bm}(t) - A(t)] \leq \mathrm{Bm}(0),$$

where the second inequality holds because $\mathrm{Bm}(t) \geq S(t)$ and $A(t) \geq 0$.

Upper bounds can now be computed by judicious choices of martingales $h(t)$ to insert in (12.5), after which the maximums on each trajectory are averaged to get the bound.

Andersen and Broadie used a martingale defined as follows. Let $L(t)$ be the discounted to time $t = 0$ of the lower bound estimate of $\mathrm{Bm}(t)$ given by the Longstaff-Schwartz method, and for $t \in \overline{T}$ set

$$h(\overline{T}_i) = h(\overline{T}_{i-1}) + L(\overline{T}_i) - L(\overline{T}_{i-1}) - \mathbb{I}_{i-1}\mathbf{E}^{(i-1)}\{L(\overline{T}_i) - L(\overline{T}_{i-1})\},$$

where $\mathbb{I}_{i-1} = 0$ if continuation is indicated at $\overline{T}_{i-1}$, and $\mathbb{I}_{i-1} = 1$ if exercise is indicated at $\overline{T}_{i-1}$.

Joshi's $h(t)$ martingale (essentially the same as the Andersen and Broadie one) is the seller's self-financing hedge consisting of one unit of Bermudan option bought at the buyer's price. The seller follows the buyer's optimal strategy giving rise to four possibilities at each exercise date: in the two cases where buyer and seller agree there is a perfect hedge; if the buyer exercises and the seller does not then the price from the optimal strategy is greater than the exercise value and the seller makes money; if the buyer does not exercise and the seller does, then the seller can re-buy the option for less than the exercise and again makes money. Spare money can be invested in the numeraire resulting in a self-financing strategy that must be a martingale after discounting.

Note that each of these techniques requires running sub-simulations on each trajectory at each reset to find either $L(\overline{T}_i)$ or $\mathrm{Bm}(\overline{T}_i)$, a total of say N_1 outer simulations, followed by say N_2 inner simulations at each reset making $N_1 \times N_2 \times M$ simulations altogether, which can evidently take up to 20 times longer than the lower bound simulation. On the other hand, variances tend to be lower cutting the number of paths needed.

12.4 Bermudan deltas

Setting $D_\ell = \frac{\partial}{\partial K(0,T_\ell)}$ as in Chapter-11, and differentiating the recursion (12.4) under the expectation with respect to the initial value $K(0, T_\ell)$ of the ℓth forward

$$D_\ell \, \mathrm{Bm}_0 = \mathbf{E} D_\ell \begin{bmatrix} \mathbb{I}[\mathcal{A}_0] S_0 + \mathbb{I}[\mathcal{A}_0^c] \mathbb{I}[\mathcal{A}_1] S_1 + \mathbb{I}[\mathcal{A}_0^c] \mathbb{I}[\mathcal{A}_1^c] \mathbb{I}[\mathcal{A}_2] S_2 + .. \\ ... + \mathbb{I}[\mathcal{A}_0^c] \mathbb{I}[\mathcal{A}_1^c] .. \mathbb{I}[\mathcal{A}_{M-2}^c] \mathbb{I}[\mathcal{A}_{M-1}] S_{M-1} \end{bmatrix},$$

$$= \mathbf{E} \left[\begin{array}{l} \mathbb{I}[\mathcal{A}_0] D_\ell S_0 + \mathbb{I}[\mathcal{A}_0^c] \mathbb{I}[\mathcal{A}_1] D_\ell S_1 \\ +.. + \mathbb{I}[\mathcal{A}_0^c] .. \mathbb{I}[\mathcal{A}_{M-2}^c] \mathbb{I}[\mathcal{A}_{M-1}] D_\ell S_{M-1} \\ + \{ D_\ell \mathbb{I}[\mathcal{A}_0] S_0 + D_\ell \mathbb{I}[\mathcal{A}_0^c] \mathbf{E}^{(0)} B_1 \} \\ + \mathbb{I}[\mathcal{A}_1^c] \{ [D_\ell \mathbb{I}[\mathcal{A}_1] S_1 + D_\ell \mathbb{I}[\mathcal{A}_1^c] \mathbf{E}^{(1)} B_2] \} \\ \vdots \\ + \mathbb{I}[\mathcal{A}_1^c] \mathbb{I}[\mathcal{A}_2^c] .. \mathbb{I}[\mathcal{A}_{M-3}^c] \left\{ \begin{array}{c} D_\ell \mathbb{I}[\mathcal{A}_{M-2}] S_{M-2} \\ + D_\ell \mathbb{I}[\mathcal{A}_{M-2}^c] \mathbf{E}^{(M-2)} B_{M-1} \end{array} \right\} \\ + \mathbb{I}[\mathcal{A}_1^c] \mathbb{I}[\mathcal{A}_2^c] .. \mathbb{I}[\mathcal{A}_{M-2}^c] \{ D_\ell \mathbb{I}[\mathcal{A}_{M-1}] S_{M-1} \} \end{array} \right]$$

and then repeatedly applying Lemma 11.1 gives

$$D_\ell B_0 = \mathbf{E} \begin{bmatrix} \mathbb{I}[\mathcal{A}_0] D_\ell S_0 + \mathbb{I}[\mathcal{A}_0^c] \mathbb{I}[\mathcal{A}_1] D_\ell S_1 + .. + \\ .. + \mathbb{I}[\mathcal{A}_0^c] .. \mathbb{I}[\mathcal{A}_{M-2}^c] \mathbb{I}[\mathcal{A}_{M-1}] D_\ell S_{M-1} \end{bmatrix}, \tag{12.7}$$

in which the $D_\ell S_i$ can be found from (9.1) and the results of Chapter-11 by partially differentiating

$$S_i = S(\overline{T}_i) = \frac{\mathrm{pSwap}(\overline{T}_i, \overline{T}_i, \overline{T}_M)}{N(\overline{T}_i)},$$

$$= Z(\overline{T}_i, \overline{T}_i) - Z(\overline{T}_i, \overline{T}_M) - \kappa \sum_{i1=i}^{M-1} \overline{\delta}_{i1} Z(\overline{T}_i, \overline{T}_{i1+1}),$$

and substituting for the $D_\ell Z(\cdot)$ and $D_\ell K(\cdot)$.

The critical simulation equations are (12.4) and (12.7). After simulating a trajectory the $\mathcal{A}_i$ (that is, the optimal stopping times) will be known from the the Longstaff-Schwartz regression technique, while the S_i and $D_\ell S_i$ being *vanilla* entities are easy to find on any trajectory. Computing B_0 and $D_\ell B_0$ is then simply a matter of averaging individual contributions over all trajectories.

Chapter 13

Vega and Shift Hedging

The *vegas* of an option, that is, its sensitivity to changes in the implied volatilities of the instruments to which the model is calibrated, are as important a risk measure as the option's deltas. But because in the shifted version of BGM, the shift $a(T)$ and volatility $\xi(t,T)$ functions are jointly fitted to swaption values during the volatility part of the calibration, the vega hedge must comprise both volatility and shift components. In this chapter we show, along the lines of Pelsser et al [89], how to compute vegas (including shift hedges) by perturbing the underlying BGM shift $a(T)$ and volatility $\xi(t,T)$ functions in such a way that only swaptions $\text{pSwpn}(t,\kappa,T_j,T_N)$ of a particular maturity T_j (but different strikes) are affected. The corresponding changes in value of the exotic option that we wish to hedge, then yield the required hedge parameters.

Denote by $\mathbb{P}$ (numeraire N_t expectation $\mathbf{E}$) whichever of the spot $\mathbb{P}_0$ or terminal $\mathbb{P}_n$ measures we are using, and let C_0 be the present value of the discounted payoff stream 'payoff $C(\cdot)$' comprising our exotic option. From Chapter-7, two swaptions say $\text{pSwpn}(t,\kappa_1,T_j,T_N)$ and $\text{pSwpn}(t,\kappa_2,T_j,T_N)$ at different strikes κ_1 and κ_2 suffice to fix first the *shift* and then the *zeta* at a particular exercise time T_j, so we assume two such swaptions will figure in the hedge. Note that these swaptions have exactly the same implied volatility and shift, differences in their values arise only from the strikes.

Perturbing by a small amount $\Delta\theta$ the shift $\alpha(T_j,T_N)$ of just the jth swaption $\text{pSwpn}(t,\kappa,T_j,T_N)$ changes its value by $\Delta_\theta\,\text{pSwpn}(0,\kappa,T_j,T_N)$ where

$$\begin{aligned}
\alpha(T_j,T_N) \quad &\rightarrow \quad (1+\Delta\theta)\,\alpha(T_j,T_N) \quad \Rightarrow \\
\Delta_\theta\,\text{pSwpn}(0,\kappa,T_j,T_N) &= \left\{ \begin{array}{c} \text{pSwpn}(0,\kappa,(1+\Delta\theta)\,\alpha(T_j,T_N)) \\ -\,\text{pSwpn}(0,\kappa,\alpha(T_j,T_N)) \end{array} \right\} \\
&= \frac{\partial}{\partial\alpha}\,\text{pSwpn}(0,\kappa,\alpha(T_j,T_N)) \;\times\; \alpha(T_j,T_N)\;\Delta\theta.
\end{aligned}$$

Similarly, perturbing by a small amount $\Delta\varepsilon$ the instantaneous swaption volatility $\sigma(t,T_j,T_N)$ of just the jth swaption changes its implied volatility $\beta(T_j,T_N)$, which in turn changes its value by $\Delta_\varepsilon\,\text{pSwpn}(0,\kappa,T_j,T_N)$ where

$$\begin{aligned}
\sigma(t,T_j,T_N) \quad &\rightarrow \quad (1+\Delta\varepsilon)\,\sigma(t,T_j,T_N) \qquad \Rightarrow \\
\beta^2(T_j,T_N)\;T_j = \int_0^{T_j} |\sigma(t,T_j,T_N)|^2\,dt \quad &\rightarrow \quad (1+\Delta\varepsilon)^2\,\beta^2(T_j,T_N)\;T_j \quad \textit{and}
\end{aligned}$$

$$\Delta_\varepsilon \,\text{pSwpn}\left(0,\kappa,T_j,T_N\right) = \begin{Bmatrix} \text{pSwpn}\left(0,\kappa,\left(1+\Delta\varepsilon\right)\beta\left(T_j,T_N\right)\right) \\ -\,\text{pSwpn}\left(0,\kappa,\beta\left(T_j,T_N\right)\right) \end{Bmatrix}$$
$$= \frac{\partial}{\partial\beta}\,\text{pSwpn}\left(0,\kappa,\beta\left(T_j,T_N\right)\right) \;\times\; \beta\left(T_j,T_N\right)\,\Delta\varepsilon.$$

The partial derivatives are easy to compute; from (4.16) and (A.2.3)

$$\text{pSwpn}\left(0,\kappa,\alpha,\beta\right) = \text{level} \times \mathbf{B}\left\{\omega\left(0\right)+\alpha,\kappa+\alpha,\beta\sqrt{T_j}\right\} \quad \Rightarrow$$
$$\frac{\partial}{\partial\beta}\,\text{pSwpn}\left(0,\kappa,\alpha,\beta\right) = \text{level} \times \left(\kappa+\alpha\right)\mathbf{N}'\left(h-\beta\sqrt{T_j}\right)\,\sqrt{T_j},$$
$$\frac{\partial}{\partial\alpha}\,\text{pSwpn}\left(0,\kappa,\alpha,\beta\right) = \text{level} \times \left\{\mathbf{N}\left(h\right)-\mathbf{N}\left(h-\beta\sqrt{T_j}\right)\right\},$$
$$\textit{where} \qquad h = \frac{\ln\frac{\omega(0)+\alpha}{\kappa+\alpha}+\frac{1}{2}\beta^2 T_j}{\beta\sqrt{T_j}}.$$

Crucially, we will construct perturbations in the BGM shift $a\left(T\right)$ and volatility $\xi\left(t,T\right)$ functions, that respectively change the shift $\alpha\left(T_j,T_N\right)$ and volatility $\sigma\left(t,T_j,T_N\right)$ of *just* the jth swaptions $\text{pSwpn}\left(0,\kappa,T_j,T_N\right)$ *and no others.* That enables us to find the changes $\Delta_\theta C_0$ and $\Delta_\varepsilon C_0$ in value of the exotic option corresponding to those perturbations in $a\left(T\right)$ and $\xi\left(t,T\right)$ respectively (on a trajectory-by-trajectory basis if pricing is by simulation), equate those changes to the corresponding swaption changes like

$$\Delta_\theta C_0 = a_1\;\Delta_\theta\,\text{pSwpn}\left(0,\kappa_1,T_j,T_N\right)+a_2\;\Delta_\theta\,\text{pSwpn}\left(0,\kappa_2,T_j,T_N\right),$$
$$\Delta_\varepsilon C_0 = a_1\;\Delta_\varepsilon\,\text{pSwpn}\left(0,\kappa_1,T_j,T_N\right)+a_2\;\Delta_\varepsilon\,\text{pSwpn}\left(0,\kappa_2,T_j,T_N\right),$$

and then solve these two equations for the *vega and shift hedge pair* $\left(a_1,a_2\right)$ of the exotic option into the *pair* of swaptions $\text{pSwpn}\left(0,\kappa_1,T_j,T_N\right)$ and $\text{pSwpn}\left(0,\kappa_2,T_j,T_N\right)$.

For clarity of exposition, we first derive the required BGM shift and volatility perturbations for coterminal swaptions and then consider the case when calibration is to a smaller miscellaneous set of liquid instruments.

13.1 When calibrated to coterminal swaptions

Assume fixed and floating nodes coincide, and there is a full complement of coterminal quarterly swaptions (with the last ones caplets) exercising at T_j for $j = 1,..,N-1$ and all maturing at T_N.

13.1.1 The shift part

Ignoring the spread μ_j, for $j = 1, 2, .., N-1$

$$\alpha(T_j, T_N) = \sum_{\ell=j}^{N-1} u_{j,\ell}^{(N)} a(T_\ell), \quad u_{j,\ell}^{(N)} = \frac{\delta_\ell B(0, T_{\ell+1})}{\sum_{k=j}^{N-1} \delta_k B(0, T_{k+1})} \tag{13.1}$$

$$\textit{that is} \qquad \underset{(N-1)\times 1}{\alpha} = \underset{(N-1)\times(N-1)}{u} \; \underset{(N-1)\times 1}{a} \qquad \textit{or}$$

$$\begin{pmatrix} \alpha(T_1, T_N) \\ \vdots \\ \alpha(T_{N-1}, T_N) \end{pmatrix} = \begin{pmatrix} u_{1,1}^{(N)} & \cdot^{\cdot^{\cdot}} & u_{1,N-1}^{(N)} \\ 0 & \ddots & \vdots \\ 0 & 0 & u_{N-1,N-1}^{(N)} \end{pmatrix} \begin{pmatrix} a(T_1) \\ \vdots \\ a(T_{N-1}) \end{pmatrix}.$$

Being a non-singular upper triangular $(N-1)\times(N-1)$ matrix, u has both a right and left inverse u^{-1} which is also $(N-1)\times(N-1)$ and upper triangular. Specifically, subtracting (13.1) at j and $(j+1)$ and simplifying

$$a(T_j) = \alpha(T_{j+1}, T_N) + \frac{1}{u_{j,\,j}^{(N)}} \left[\alpha(T_j, T_N) - \alpha(T_{j+1}, T_N)\right] \qquad \Rightarrow$$

$$a \;=\; u^{-1}\,\alpha \qquad \textit{where}$$

$$u^{-1} = \begin{pmatrix} \frac{1}{u_{1,1}^{(N)}} & 1 - \frac{1}{u_{1,1}^{(N)}} & 0 & \ddots & 0 \\ 0 & \ddots & \ddots & \ddots & \vdots \\ 0 & \cdots & \frac{1}{u_{j,\,j}^{(N)}} & 1 - \frac{1}{u_{j,\,j}^{(N)}} & 0 \\ 0 & \cdots & \cdots & \ddots & \vdots \\ 0 & \cdots & \cdots & 0 & \frac{1}{u_{N-1,N-1}^{(N)}} \end{pmatrix}.$$

Now proportionally perturb just the jth swaption shift by an amount $\Delta\theta$

$$\begin{aligned} \alpha(T_j, T_N) \quad &\rightarrow \quad (1 + \Delta\theta)\, \alpha(T_j, T_N) \\ \textit{that is} \quad \alpha \quad &\rightarrow \quad \alpha + \Delta\theta \; \left(0, \ldots, \alpha(T_j, T_N), \ldots, 0\right)^T. \end{aligned}$$

The corresponding perturbation in the BGM shift function that changes the shift $\alpha(T_j, T_N)$ in just the jth swaption and no others must be

$$\begin{aligned} a &\rightarrow u^{-1} \left\{\alpha + \Delta\theta \; \left(0, \ldots, \alpha(T_j, T_N), \ldots, 0\right)^T\right\}, \\ &= a + \Delta\theta \; u^{-1} \left(0, \ldots, \textstyle\sum_{\ell=j}^{N-1} u_{j,\ell}^{(N)} a(T_j), \ldots, 0\right)^T. \end{aligned}$$

Note that this perturbation affects just $a(T_{j-1})$ and $a(T_j)$ with all other $a(T_j)$ remaining unchanged

$$a(T_{j-1}) \rightarrow a(T_{j-1}) + \Delta\theta \left(1 - \frac{1}{u^{(N)}_{j-1,\ j-1}}\right) \sum_{\ell=j}^{N-1} u^{(N)}_{j,\ \ell} a(T_\ell), \qquad (13.2)$$

$$a(T_j) \rightarrow a(t, T_j) + \Delta\theta \frac{1}{u^{(N)}_{j,\ j}} \sum_{\ell=j}^{N-1} u^{(N)}_{j,\ \ell} a(T_\ell).$$

REMARK 13.1 Given the orders of magnitude of $u^{(N)}_{j,\ell}$ (roughly $\frac{1}{N-j}$) clearly the numbers $\frac{1}{u^{(N)}_{j,\ j}}$ can be quite large, but will be kept under control so long as swaption shifts $\alpha(T_j, T_N)$ are not changing rapidly from tenor T_j to tenor T_{j+1}. Similarly for the $A^{(N)}_{j,\ell}$ appearing in the volatility part below, where the swaption volatility $\sigma(t, T_j, T_N)$ must be stable from tenor to tenor. $\square$

13.1.2 The volatility part

Construction of the volatility perturbation must allow for $\xi(t, T)$ being vector valued and time dependent (unlike the shift $a(T)$); hence the following *die-at-exercise convention*:

Condition 1 *Forward and swaprate volatilities satisfy*

$$\xi(t, T_j) = 0 \quad \textit{and} \quad \sigma(t, T_j, T_N) = 0 \quad \textit{for} \quad t > T_j. \qquad (13.3)$$

From (4.16) with weights $A^{(N)}_{j,\ell}$ that depend only on the initial yieldcurve and assuming k factors, the volatilities of a set of coterminal swaptions maturing at T_N can be written (the range of σ and ξ must be k-dimensional) for $j = 1, 2, .., N-1$ as

$$\sigma(t, T_j, T_N) = \sum_{\ell=j}^{N-1} A^{(N)}_{j,\ell} \xi(t, T_j), \qquad (13.4)$$

$$\textit{that is} \quad \underset{(N-1)\times k}{\sigma(t)} = \underset{(N-1)\times(N-1)}{A} \ \underset{(N-1)\times k}{\xi(t)} \quad \textit{or}$$

$$\begin{pmatrix} \sigma(t, T_1, T_N) \\ \vdots \\ \sigma(t, T_{N-1}, T_N) \end{pmatrix} = \begin{pmatrix} A^{(N)}_{1,1} & \ddots & A^{(N)}_{1,N-1} \\ 0 & \ddots & \vdots \\ 0 & 0 & A^{(N)}_{N-1,N-1} \end{pmatrix} \begin{pmatrix} \xi(t, T_1) \\ \vdots \\ \xi(t, T_{N-1}) \end{pmatrix}.$$

Being a non-singular upper triangular $(N-1)\times(N-1)$ matrix, A has both a right and left inverse A^{-1} which is also $(N-1)\times(N-1)$ and upper triangular, and therefore

$$\xi(t) = A^{-1}\sigma(t).$$

REMARK 13.2 The equations $\sigma(t) = A\xi(t)$ and $\xi(t) = A^{-1}\sigma(t)$ must hold for all times $t \in [0, T_N]$, during which time the component swaptions are consecutively exercising. The time convention Condition-13.3 ensures there is no discrepancy. For example, deleting the 1st row from A and 1st column from A^{-1} is equivalent to eliminating the first calibration instrument:

$$\begin{aligned} &A \to A(2:N\,,\,1:N) \qquad A^{-1} \to A^{-1}(1:N\,,\,2:N), \\ &\Rightarrow A(2:N\,,\,1:N) * A^{-1}(1:N\,,\,2:N) = I_{N-2}, \end{aligned}$$

because of the upper triangular nature of A. Then time can be moved forward a quarter by deleting a column in A and row in A^{-1}:

$$\begin{aligned} &A(2:N\,,\,1:N) \to A(2:N\,,\,2:N) \quad \textit{and} \\ &A^{-1}(1:N\,,\,2:N) \to A^{-1}(2:N\,,\,2:N), \\ \Rightarrow\quad &A(2:N\,,\,2:N) * A^{-1}(2:N\,,\,2:N) = I_{N-2}. \end{aligned}$$

The point is that it does not matter whether we work with the full $(N-1) \times (N-1)$ matrix A or its submatrices; due to the triangular nature of A there is no inconsistency in ignoring the first instrument according to Condition-13.3 once it has matured. ∎

In the case of the *usual simple approximation* to the $A^{(N)}_{j,\ell}$

$$A^{(N)}_{j,\ell} = \begin{cases} 0 \quad \textit{for} \quad \ell < j \\ \dfrac{\delta_\ell B(0, T_{\ell+1}) H(0, T_\ell)}{\sum_{k=j}^{N-1} \delta_k B(0, T_{k+1}) H(0, T_k)} \end{cases} \tag{13.5}$$

the inverse A^{-1} can be computed like u^{-1} for the shift. Subtracting (13.4) at j and $(j+1)$ and simplifying

$$\xi(t, T_j) = \sigma(t, T_{j+1}, T_N) + \frac{1}{A^{(N)}_{j,\,j}}\left[\sigma(t, T_j, T_N) - \sigma(t, T_{j+1}, T_N)\right] \qquad \Rightarrow$$

$$\xi \;=\; A^{-1}\sigma \qquad \textit{where}$$

$$A^{-1} = \begin{pmatrix} \frac{1}{A^{(N)}_{1,1}} & 1 - \frac{1}{A^{(N)}_{1,1}} & 0 & \ddots & 0 \\ 0 & \ddots & \ddots & \ddots & \vdots \\ 0 & \dots & \frac{1}{A^{(N)}_{j,\,j}} & 1 - \frac{1}{A^{(N)}_{j,\,j}} & 0 \\ 0 & \dots & \dots & \ddots & \vdots \\ 0 & \dots & \dots & 0 & \frac{1}{A^{(N)}_{N-1,N-1}} \end{pmatrix}.$$

Now proportionally perturb just the jth swaprate volatility by an amount $\Delta\varepsilon$

$$\begin{aligned} \sigma(t, T_j, T_N) &\rightarrow (1 + \Delta\varepsilon)\, \sigma(t, T_j, T_N) \\ \textit{that is} \quad \sigma &\rightarrow \sigma + \Delta\varepsilon \, \left(0, \ldots, \sigma(t, T_j, T_N), \ldots, 0\right)^T \end{aligned}$$

in a scalar fashion that importantly leaves swaprate correlation unaltered. The perturbation in the BGM volatility function corresponding to this change in just the jth swaprate volatility $\sigma(t, T_j, T_N)$ (and no others) is given by

$$\begin{aligned} \xi &\rightarrow A^{-1} \left\{\sigma + \Delta\varepsilon \, \left(0, \ldots, \sigma(t, T_j, T_N), \ldots, 0\right)^T\right\}, \\ &= \xi + \Delta\varepsilon \; A^{-1} \left(0, \ldots, \textstyle\sum_{\ell=j}^{N-1} A_{j,\ell}^{(N)} \xi(t, T_j), \ldots, 0\right)^T . \end{aligned}$$

In the case of the *usual simple approximation* (13.5), the perturbation will affect just $\xi(t, T_{j-1})$ and $\xi(t, T_j)$ with all other $\xi(t, T_j)$ remaining unchanged

$$\begin{aligned} \xi(t, T_{j-1}) &\rightarrow \xi(t, T_{j-1}) + \Delta\varepsilon \left(1 - \frac{1}{A_{j-1,\, j-1}^{(N)}}\right) \sum_{\ell=j}^{N-1} A_{j,\,\ell}^{(N)} \xi(t, T_\ell), \\ \xi(t, T_j) &\rightarrow \xi(t, T_j) + \Delta\varepsilon \frac{1}{A_{j,\, j}^{(N)}} \sum_{\ell=j}^{N-1} A_{j,\,\ell}^{(N)} \xi(t, T_\ell). \end{aligned} \tag{13.6}$$

REMARK 13.3 Problems may also arise in practice if N and hence the reciprocals of $A_{j,\, j}^{(N)}$ are large and concurrently the relative perturbation $\Delta\varepsilon$ is substantial to get distinct changes in value under simulation. □

REMARK 13.4 Together (13.2) and (13.6) give some idea of the kinds of perturbation of the BGM shift $a(t)$ and volatility $\xi(t, T)$ functions needed to produce changes in one particular swaption, namely joint changes in just two adjacent (by maturity) values of $\xi(t, T)$. *Bump and grind* methods for computing vegas involve a recalibration that must try to mimic these perturbations in some way; that is not easy, and leads to inaccuracies when the functions are overly distorted. □

13.2 When calibrated to liquid swaptions

Assume calibration is to a liquid set of m swaptions and caplets exercising at different times T_j and with mixed maturities T_n where $n \leq N$. The matrices u and A in (13.1) and (13.4) actually specify the calibration instruments, so if we make the convention that the $u_{j,\ell}^{(N)}$ and $A_{j,\ell}^{(N)}$ for our m calibration instruments are entered in rows by exercise, then we can talk without ambiguity about *the calibration* $\langle u, A\rangle$. For N nodes, u and A will therefore be of order $m\times(N-1)$ with $m \leq N-1$, and also *block triangular*

$$\begin{aligned} &X_{j,1} = X_{j,2} = ... = X_{j,\ell-1} = 0 \quad \textit{and} \quad X_{j,\ell} \neq 0 \\ \Rightarrow \quad &X_{j1,\ell 1} = 0 \quad \textit{for} \quad j1 > j \quad \textit{and} \quad \ell 1 < \ell. \end{aligned}$$

Deleting a row removes a calibration instrument, while deleting columns of zeroes at the front of the matrix corresponds to moving forward in time.

Practically speaking, with quarterly nodes N will be over 100 and m much less. Moreover, we can expect $\operatorname{rank} u = \operatorname{rank} A = m$ (otherwise the linear dependence indicates an unsatisfactory calibration set) so right inverses u^{-1} and A^{-1} will exist (see Section-A.4.2) such that

$$\begin{aligned} a = u^{-1}\,\alpha \quad &\textit{solves} \quad \alpha = u\ a \quad \textit{with} \quad \|a\|_2 \quad \textit{minimal,} \\ \textit{and} \quad \xi = A^{-1}\,\sigma \quad &\textit{solves} \quad \sigma = A\ \xi \quad \textit{with each} \quad \left\|\xi^{(k)}\right\|_2 \quad \textit{minimal.} \end{aligned}$$

Similarly to Section-13.1, perturbations in the shift and volatility of the jth instrument

$$\begin{aligned} \alpha \quad &\rightarrow \quad \alpha + \Delta\theta\ \left(0, \ldots, \alpha\left(T_j, T_N\right), \ldots, 0\right)^T, \\ \sigma \quad &\rightarrow \quad \sigma + \Delta\varepsilon\ \left(0, \ldots, \sigma\left(t, T_j, T_N\right), \ldots, 0\right)^T, \end{aligned}$$

are induced by the corresponding perturbations

$$\begin{aligned} a &\rightarrow u^{-1}\left\{\alpha + \Delta\theta\ \left(0, \ldots, \alpha\left(T_j, T_N\right), \ldots, 0\right)^T\right\}, \\ \xi &\rightarrow A^{-1}\left\{\sigma + \Delta\varepsilon\ \left(0, \ldots, \sigma\left(t, T_j, T_N\right), \ldots, 0\right)^T\right\}, \end{aligned}$$

in the BGM shift and volatility functions.

REMARK 13.5 As in Section-13.1, the volatility perturbation is consistent with the time convention (13.3), because if we strike calibration instruments as they exercise along with columns corresponding to the past, the resulting submatrices will still be mutually inverse though no longer yielding an ξ with minimal 2-norm. Alternatively, to ensure an ξ with minimal 2-norm, we could recompute ξ for each internodal period working with consecutive matrices of decreasing size as instruments exercise and time moves forward. □

REMARK 13.6 A different approach to finding the required perturbations when calibration is to liquid instruments, is to use linear programming with extra constraints designed to produce a stable and accurate result. □

Chapter 14

Cross-Economy BGM

The cross-economy version of shifted BGM links foreign and domestic shifted BGM models via the forward FX exchange rate. In the Gaussian HJM framework the link is possible with deterministic volatilities for domestic and foreign instantaneous forwards, and also the FX rate; that is, deterministic volatilities are totally compatible with lognormal models for the prices of domestic and foreign bonds and the spot and forward FX rates.

But, as shown by Schlogl [112], in cross-economy BGM some among the domestic and foreign interest rate volatilities, and FX forward volatilities must be stochastic. Nevertheless, as with swaption volatilities in domestic BGM, with appropriate choices it is possible to obtain approximations for stochastic volatilities in cross-economy BGM that are good enough to return by simulation fairly accurate values for the implied volatilities to which the model is calibrated.

To set the scene, in the following Section-14.1 we first work through relevant ideas in a cross-economy HJM framework, before considering the BGM equivalent. Our notation for the foreign economy will be to superfix f to domestic variables to denote the equivalent foreign variable; for example, if $\mathbb{P}_T$ and $B(t,T)$ are respectively the T-forward measure and zero coupon in the domestic economy, then $\mathbb{P}_T^f$ and $B^f(t,T)$ are the equivalent in the foreign economy.

14.1 Cross-economy HJM

The extra *ingredients* needed in HJM to cope with a foreign economy are:

- a *spot FX rate* $S(t)$, which is the price of one *foreign zlotty* in *domestic dollars* $Ƶ1 = \$S(t)$, with volatility function $\nu(t)$, drift $\alpha^S(t)$ and SDE

$$dS(t) = S(t)\left[\alpha^S(t)\,dt + \nu^*(t)\,dW_0(t)\right],$$

- an instantaneous *foreign forward rate* $f^f(t,T)$ at time t for maturity T with volatility function $\sigma^f(t,T)$, drift $\alpha^f(t,T)$ and SDE

$$df^f(t,T) = \alpha^f(t,T)\,dt + \sigma^{f*}(t,T)\,dW_0(t) \tag{14.1}$$

- a *foreign spot rate* $r^f(t)$ and *foreign bank account* $\beta^f(t)$ defined by

$$r^f(t) = f^f(t,t), \qquad \beta^f(t) = \exp\left(\int_0^t r^f(s)\,ds\right),$$

- *foreign zero coupon bonds* $B^f(t,T)$ maturing at T defined by

$$B^f(t,T) = \exp\left(-\int_t^T f^f(t,u)\,du\right). \tag{14.2}$$

The FX drift $\alpha^S(t)$ is fixed by the fact that $\mathcal{Z}1$ invested in the foreign bank account and converted to domestic dollars constitutes a domestic asset, and so its discounted value must be a $\mathbb{P}_0$-martingale; that is, the drift in

$$\begin{aligned}\frac{d\left(\frac{\beta^f(t)\,S(t)}{\beta(t)}\right)}{\frac{\beta^f(t)\,S(t)}{\beta(t)}} &= \left[\alpha^S(t)\,dt + \nu^*(t)\,dW_0(t)\right] + r^f(t)\,dt - r(t)\,dt,\\ &= \left[\alpha^S(t) + r^f(t) - r(t)\right]dt + \nu^*(t)\,dW_0(t)\end{aligned}$$

must be zero, giving $\alpha^S(t) = r(t) - r^f(t)$. Hence the SDE for $S(t)$ is

$$\frac{dS(t)}{S(t)} = \left[r(t) - r^f(t)\right]dt + \nu^*(t)\,dW_0(t). \tag{14.3}$$

Similarly the forward drifts $\alpha^f(t,T)$ are fixed by the fact that all foreign zero coupon bonds converted to dollars are domestic assets, and so their discounted values must be $\mathbb{P}_0$-martingales. Because

$$d\int_t^T f^f(t,u)\,du = \left\{\begin{array}{c} -r^f(t)\,dt + \left(\int_t^T \alpha^f(t,u)\,du\right)dt \\ +\left(\int_t^T \sigma^{f*}(t,u)\,du\right)dW_0(t)\end{array}\right\},$$

from (14.1), (14.2), (14.3) and using Ito

$$\begin{aligned}\frac{d\left(\frac{B^f(t,T)\,S(t)}{\beta(t)}\right)}{\frac{B^f(t,T)\,S(t)}{\beta(t)}} &= \left\{\left[r(t) - r^f(t)\right]dt + \nu^*(t)\,dW_0(t)\right\} - r(t)\,dt\\ &+\left\{r^f(t)\,dt - \left(\int_t^T \alpha^f(t,u)\,du\right)dt - \left(\int_t^T \sigma^{f*}(t,u)\,du\right)dW_0(t)\right\}\\ &+\frac{1}{2}\left|\int_t^T \sigma^f(t,u)\,du\right|^2 dt - \nu^*(t)\left(\int_t^T \sigma(t,u)\,du\right)dt,\end{aligned}$$

and for the drift in this SDE to be zero

$$\int_t^T \alpha^f(t,u)\,du = \frac{1}{2}\left|\int_t^T \sigma^f(t,u)\,du\right|^2 - \nu^*(t)\int_t^T \sigma^f(t,u)\,du$$

$$\Rightarrow \qquad \alpha^f(t,T) = \sigma^{f*}(t,T)\left\{\int_t^T \sigma^f(t,u)\,du - \nu(t)\right\}.$$

Hence SDEs for the foreign forwards and zero coupons are

$$df^f(t,T) = \sigma^f(t,T)\int_t^T \sigma^f(t,u)\,du\,dt + \sigma^{f*}(t,T)\left(dW_0(t) - \nu(t)\,dt\right),$$

$$\frac{dB^f(t,T)}{B^f(t,T)} = r^f(t)\,dt - \left(\int_t^T \sigma^{f*}(t,u)\,du\right)\left(dW_0(t) - \nu(t)\,dt\right), \qquad (14.4)$$

and letting $W_0^f(t)$ be Brownian motion under $\mathbb{P}_0^f$, these two expressions show the *foreign arbitrage free measure* $\mathbb{P}_0^f$ is related to $\mathbb{P}_0$ by

$$dW_0^f(t) = dW_0(t) - \nu(t)\,dt \quad \Rightarrow \quad \mathbb{P}_0^f = \mathcal{E}\left(\int_0^t \nu^*(s)\,dW_0(s)\right)\mathbb{P}_0. \quad (14.5)$$

14.2 Forward FX contracts

The value in the domestic economy of a *forward contract* $G_T(t,T_1)$ maturing at T, on a foreign zero $B^f(t,T_1)$ maturing at $T_1 = T + \delta$, must satisfy

$$\mathbf{E}_T\left\{\left[G_T(t,T_1) - B^f(T,T_1)\,S(T)\right]\middle|\,\mathcal{F}_t\right\} = 0,$$

which implies $G_T(t,T_1)$ is a $\mathbb{P}_T$-martingale with time t value

$$G_T(t,T_1) = \mathbf{E}_T\left\{\frac{B^f(T,T_1)}{B(T,T)}S(T)\middle|\,\mathcal{F}_t\right\} = \frac{B^f(t,T_1)}{B(t,T)}S(t).$$

When $T_1 = T$ the payoff from $B^f(t,T)$ at T is one zlotty $\text{Z}1 = \$\,S(T)$ and $G_T(t,T)$ becomes the *FX forward contract*

$$S_T(t) = \frac{B^f(t,T)\,S(t)}{B(t,T)} \qquad (14.6)$$

This is the well known *interest rate parity* relationship; for example, if AUS short interest rates rise ($B^f(t,T)$ gets smaller) and US rates and forward FX are unchanged ($B(t,T)$ and $S_T(t)$ remain fixed), then $S(t)$ must increase to compensate (the AUS strengthens against the USD).

Because $S_T(t)$ is a strictly positive $\mathbb{P}_T$-martingale it will have an SDE of form

$$\frac{dS_T(t)}{S_T(t)} = \nu_T^*(t)\, dW_T(t),$$

where $\nu_T(t)$ is the *instantaneous forward FX volatility.*

REMARK 14.1 It is important to appreciate that so far the relationships derived in this section are completely model free. □

14.2.1 In the HJM framework

The interest rate parity relationship (14.6) holds for all t, so on differentiating it using the HJM version of the component SDEs (1.4), (14.3), (14.4) and (14.6)

$$\frac{dB(t,T)}{B(t,T)} = r(t)\, dt - \left(\int_t^T \sigma^*(t,u)\, du\right) dW_0(t),$$

$$\frac{dS(t)}{S(t)} = \left[r(t) - r^f(t)\right] dt + \nu^*(t)\, dW_0(t), \quad \frac{dS_T(t)}{S_T(t)} = \nu_T^*(t)\, dW_T(t),$$

$$\frac{dB^f(t,T)}{B^f(t,T)} = r^f(t)\, dt - \left(\int_t^T \sigma^{f*}(t,u)\, du\right) \left(dW_0(t) - \nu(t)\, dt\right),$$

the stochastic components on both sides must tally, giving the HJM *volatility parity* relationship

$$\nu_T(t) = \nu(t) + \int_t^T \sigma(t,u)\, du - \int_t^T \sigma^f(t,u)\, du. \tag{14.7}$$

Note that at maturity T, the spot and forward FX volatilities converge, that is

$$\nu(T) = \nu_T(T).$$

Evaluating the spot volatility parity (14.7) relationship at two maturities T and T_1 ($> T$) and subtracting gives the *forward volatility parity* relationship

$$\nu_{T_1}(t) = \nu_T(t) + \int_T^{T_1} \sigma(t,u)\, du - \int_T^{T_1} \sigma^f(t,u)\, du.$$

REMARK 14.2 In the cross-economy version of Ho & Lee, which can be surprisingly useful for work on portfolios of underlying instruments like bonds (that is, not primarily volatility dependent instruments like options) both domestic and foreign interest rate volatilities are constant at say σ and σ^f respectively. Then, various possibilities for the parity relationship include:

the spot volatility $\nu(t)$ is constant and the forward volatilities $\nu_T(t)$ are time dependent

$$\nu(t) = \nu \quad \Rightarrow \quad \nu_T(t) = \nu + \left(\sigma - \sigma^f\right)(T - t);$$

the forward FX volatilities $\nu_T(t)$ are dependent only on maturity T but the spot volatility is time t dependent

$$\nu_T(t) = \nu + \left(\sigma - \sigma^f\right)T \quad \Rightarrow \quad \nu(t) = \nu + \left(\sigma - \sigma^f\right)t.$$

□

14.2.2 In the BGM framework

We now develop the BGM equivalents of the cross-economy HJM measure change (14.5), and the HJM volatility parity relationship (14.7). Dividing the (model free) spot parity relationship (14.6) at the two maturities T and $T_1\ (> T)$ gives a forward version of the parity relationship

$$F_T^f(t, T_1) = \frac{S_{T_1}(t)}{S_T(t)} F_T(t, T_1). \tag{14.8}$$

Note that in this equation $F_T^f(t, T_1)$ is a T-maturing forward contract on the zero $B^f(t, T)$ all within the foreign economy, and is quite different to $G_T(t, T_1)$. This equation holds for all t, so matching SDEs on each side must yield the required measure changes and volatility parity relationships between the two economies.

Under the T-forward measures $\mathbb{P}_T$ and $\mathbb{P}_T^f$, SDEs for the components are

$$\frac{dF_T^f(t, T_1)}{F_T^f(t, T_1)} = -b^{f*}(t, T, T_1)\, dW_T^f(t), \tag{14.9}$$

$$\frac{dF_T(t, T_1)}{F_T(t, T_1)} = -b^*(t, T, T_1)\, dW_T(t), \qquad \frac{dS_T(t)}{S_T(t)} = \nu_T^*(t)\, dW_T(t),$$

$$\frac{dS_{T_1}(t)}{S_{T_1}(t)} = \nu_{T_1}^*(t)\left[b(t, T, T_1)\, dt + dW_T(t)\right],$$

from which an SDE for the right-hand side of (14.8) is

$$\frac{d\left(\frac{S_{T_1}(t)}{S_T(t)} F_T(t, T_1)\right)}{\left(\frac{S_{T_1}(t)}{S_T(t)} F_T(t, T_1)\right)} = -\left[b(t, T, T_1) - \nu_{T_1}(t) + \nu_T(t)\right]^* \left[dW_T(t) - \nu_T(t)\, dt\right].$$

Matching this SDE to that of the foreign forward contract $F_T^f(t, T_1)$ gives

$$\left[b(t, T, T_1) - \nu_{T_1}(t) + \nu_T(t)\right]^* \left[dW_T(t) - \nu_T(t)\, dt\right] = b^{f*}(t, T, T_1)\, dW_T^f(t).$$

That means that the measure change between the domestic T-forward measure $\mathbb{P}_T$ and foreign T-forward measure $\mathbb{P}_T^f$ must be determined by

$$dW_T^f(t) = dW_T(t) - \nu_T(t)\,dt \Rightarrow \mathbb{P}_T^f = \mathcal{E}\left(\int_0^t \nu_T^*(s)\,dW_T(s)\right)\mathbb{P}_T, \quad (14.10)$$

and the *BGM volatility parity* relationship connecting FX forward volatilities and domestic and foreign bond volatility differences is

$$\begin{aligned} \nu_{T_1}(t) - \nu_T(t) &= b(t,T,T_1) - b^f(t,T,T_1), \quad (14.11)\\ &= \frac{\delta H(t,T)}{[1+\delta K(t,T)]}\xi(t,T_j) - \frac{\delta H^f(t,T)}{[1+\delta K^f(t,T)]}\xi^f(t,T_j). \end{aligned}$$

The *change of measure* relationship (14.10) with maturity T set at the end of the current interval $T = T_@$, also relates the spot measures in the domestic and foreign economies

$$dW_0^f(t) = dW_0(t) - \nu_@(t)\,dt \quad (14.12)$$

where $\nu_@(t)$ is the instantaneous volatility of the immediately maturing FX forward contract $S_{T_@}(t) = S_@(t)$ which has SDE

$$\frac{dS_@(t)}{S_@(t)} = \nu_@^*(t)\,dW_0(t). \quad (14.13)$$

REMARK 14.3 A version of shifted cross-economy BGM is possible with the FX also having a term structure of shift $f(T)$. Suppose $T_1 = T + \delta$, relevant SDEs are then the SDEs for $F_T^f(t,T_1)$ and $F_T(t,T_1)$ as in (14.9), with the two new SDEs for $S_T(t)$ and $S_{T_1}(t)$

$$\begin{aligned} \frac{dS_T(t)}{S_T(t)} &= \frac{S_T(t)+f(T)}{S_T(t)}\nu_T^*(t)\,dW_T(t),\\ \frac{dS_{T_1}(t)}{S_{T_1}(t)} &= \frac{S_{T_1}(t)+f(T_1)}{S_{T_1}(t)}\nu_{T_1}^*(t)\left[b(t,T,T_1)\,dt + dW_T(t)\right]. \end{aligned}$$

Similarly to (14.11), the volatility parity relationship becomes

$$\begin{aligned} &\frac{S_{T_1}(t)+f(T_1)}{S_{T_1}(t)}\nu_{T_1}(t) - \frac{S_T(t)+f(T)}{S_T(t)}\nu_T(t)\\ &= \frac{\delta[K(t,T)+a(T)]}{[1+\delta K(t,T)]}\xi(t,T) - \frac{\delta\left[K^f(t,T)+a^f(T)\right]}{[1+\delta K^f(t,T)]}\xi^f(t,T), \end{aligned}$$

creating a problem of how to handle the stochastic parts of the FX volatility components! □

14.3 Cross-economy models

Assume for simplicity that fixed and floating nodes T_j $(j = 0, 1, 2, ...)$ coincide in the two economies. The volatility parity equation (14.11) for successive nodes T_j and T_{j+1} is

$$\nu_{T_{j+1}}(t) = \nu_{T_j}(t) + b(t, T_j, T_{j+1}) - b^f(t, T_j, T_{j+1}).$$

Then, substituting for the bond volatilities from (3.3)

$$b(t, T_j, T_{j+1}) = \frac{\delta_j H(t, T_j)}{[1 + \delta_j K(t, T_j)]} \xi(t, T_j) = h_j(t) \xi(t, T_j),$$

$$b^f(t, T_j, T_{j+1}) = \frac{\delta_j H^f(t, T_j)}{[1 + \delta_j K^f(t, T_j)]} \xi^f(t, T_j) = h_j^f(t) \xi^f(t, T_j),$$

we obtain a vector system of *forward volatility parity equations* for $j = 1, 2, ...$

$$\nu_{T_{j+1}}(t) = \nu_{T_j}(t) + h_j(t)\, \xi(t, T_j) - h_j^f(t)\, \xi^f(t, T_j). \tag{14.14}$$

Within this system of equations the $h_j(\cdot)$ and $h_j^f(\cdot)$ are stochastic, and so unlike HJM, it is not possible to concurrently have domestic rates modelled by shifted BGM (that is $\xi(t, T)$ deterministic), foreign rates also modelled by shifted BGM (that is $\xi^f(t, T)$ deterministic), and FX forward volatilities lognormal (that is $\nu_T(t)$ deterministic).

Practically useful possibilities reduce to one of:

1. Domestic and foreign rates are shifted BGM and just one of the FX forward volatilities, say $\nu_{T_k}(t)$, is deterministic. In that case, the parity equations (14.14) say all other forward volatilities $\nu_{T_j}(t)$ $(j \neq k)$ must be stochastic. Note that this includes the possibility, see Section-14.4 below, of the chosen deterministic forward volatility being at one maturity on some time intervals and other maturities at other times.

2. Domestic rates are shifted BGM and all FX forward volatilities are deterministic, which means foreign rates cannot be shifted BGM and foreign shifted forward volatilities $\xi^f(t, T_j)$ must be stochastic.

3. Foreign rates are shifted BGM and all FX forward volatilities are deterministic, so domestic rates are not shifted BGM and domestic shifted forward volatilities $\xi(t, T_j)$ are stochastic.

4. Recall that $h_j^f(t)$ and $h_j(t)$ $(j = 1, 2, ..)$ are low variance martingales, and use the deterministic approximation

$$\nu_{T_{j+1}}(t) \cong \nu_{T_j}(t) + h_j(0)\, \xi(t, T_j) - h_j^f(0)\, \xi^f(t, T_j) \tag{14.15}$$

in a setup which assumes domestic and foreign rates are shifted BGM and all FX forward volatilities are deterministic.

Often the choice will be suggested by the instrument to be valued, with the change of measure formula (14.10) allowing stochastic variables in both domestic and foreign economies to be expressed in terms of measures convenient for computation. In the next example, for instance, the best choice is clearly foreign rates and FX deterministic.

Example 14.1
The T-maturity roll $Q(t,T)$ of a *quanto* paying foreign Libor in domestic dollars at T_1 will have time-t value

$$
\begin{aligned}
Q(t,T) &= B(t,T_1)\,\mathbf{E}_{T_1}\left\{\delta K^f(T,T)\middle|\,\mathcal{F}_t\right\},\\
&= \delta B(t,T_1)\,\mathbf{E}_{T_1}\left\{H^f(T,T) - a^f(T)\middle|\,\mathcal{F}_t\right\},\\
&= \delta B(t,T_1)\,\mathbf{E}_{T_1}\left\{H^f(t,T)\,\mathcal{E}\left(\int_t^T \xi^{f*}(s,T)\,dW^f_{T_1}(s)\right) - a^f(T)\middle|\,\mathcal{F}_t\right\},\\
&= \delta B(t,T_1)\,\mathbf{E}_{T_1}\left\{\begin{array}{c} H^f(t,T)\,\mathcal{E}\left(\int_t^T \xi^{f*}(s,T)\left[dW_{T_1}(s) - \nu_{T_1}(s)\,ds\right]\right)\\ -a^f(T)\end{array}\middle|\,\mathcal{F}_t\right\},\\
&= \delta B(t,T_1)\left\{\begin{array}{c}\left[K^f(t,T) + a(T)\right]\exp\left(-\int_t^T \xi^{f*}(s,T)\,\nu_{T_1}(s)\,ds\right)\\ -a^f(T)\end{array}\right\},
\end{aligned}
$$

if $\xi^f(s,T)$ and $\nu_{T_1}(s)$ are taken to be deterministic. ☐

14.4 Model with the spot volatility deterministic

In order to both approximately fit the implied volatilities of forward FX options and be able to simulate accurately, we will assume that:

1. Both domestic and foreign forward rates are shifted lognormal, that is both $\xi(t,T_j)$ and $\xi^f(t,T_j)$ are deterministic.

2. On consecutive intervals the instantaneous forward volatility $\nu_{T_@}(t) = \nu_@(t)$ of the next maturing forward FX contract $S_{T_@}(t) = S_@(t)$ is deterministic.

From (14.14), these two assumptions mean FX forward volatilities away from the current interval ought to be stochastic, but we will use (14.15) to approximate them deterministically. The implied volatility $\alpha(T_j)$ of the T_j exercising forward FX option is

$$
\alpha^2(T_j)\,T_j = \int_0^{T_j} \left|\nu_{T_j}(t)\right|^2 dt = \sum_{k=1}^{j} \int_{T_{k-1}}^{T_k} \left|\nu_{T_j}(t)\right|^2 dt, \tag{14.16}
$$

in which the forward volatilities $\nu_{T_j}(t)$ on $t \in (T_{k-1}, T_k]$ are computed from the following table

k	$t \in (T_{k-1}, T_k]$	$\nu_{T_j}(t) =$
1	$(T_0, T_1]$	$\nu_{T_1}(t) + b(t, T_1, T_j) - b^f(t, T_1, T_j)$
2	$(T_1, T_2]$	$\nu_{T_2}(t) + b(t, T_2, T_j) - b^f(t, T_2, T_j)$
3	$(T_2, T_3]$	$\nu_{T_3}(t) + b(t, T_3, T_j) - b^f(t, T_3, T_j)$
$\vdots$	$\vdots$	$\vdots$
$j-1$	$(T_{j-2}, T_{j-1}]$	$\nu_{T_{j-1}}(t) + b(t, T_{j-1}, T_j) - b^f(t, T_{j-1}, T_j)$
j	$(T_{j-1}, T_j]$	$\nu_{T_j}(t)$

with the domestic $b(t, T_{j1}, T_{j2})$ and foreign $b^f(t, T_{j1}, T_{j2})$ bond volatility differences deterministically approximated by repeatedly applying (3.3) and setting stochastic parts to their initial value

$$b(t, T_k, T_j) \cong \sum_{\ell=k}^{j-1} \frac{\delta_\ell H(0, T_\ell)}{[1 + \delta_\ell K(0, T_\ell)]} \xi(t, T_\ell) = \sum_{\ell=k}^{j-1} h_\ell(0)\, \xi(t, T_\ell), \quad (14.17)$$

$$b^f(t, T_k, T_j) \cong \sum_{\ell=k}^{j-1} \frac{\delta_\ell H^f(0, T_\ell)}{[1 + \delta_\ell K^f(0, T_\ell)]} \xi^f(t, T_\ell) = \sum_{\ell=k}^{j-1} h_\ell^f(0)\, \xi^f(t, T_\ell).$$

Putting it all together

$$\alpha^2(T_j)\ T_j = \sum_{k=1}^{j} \int_{T_{k-1}}^{T_k} \left| \begin{array}{c} \nu_{T_k}(t) + \sum_{\ell=k}^{j-1} h_\ell(0)\, \xi(t, T_\ell) \\ - \sum_{\ell=k}^{j-1} h_\ell^f(0)\, \xi^f(t, T_\ell) \end{array} \right|^2 dt \quad (14.18)$$

REMARK 14.4 In the Pedersen cross-economy calibration scheme below equation (14.18) will connect the varying trial values for $\xi(t, T_j)$, $\xi^f(t, T_j)$ and $\nu_@(t)$ with the target implied volatilities $\alpha(T_j)$. □

Simulation can now be done under the spot measure using virtually any scheme. For example, with the notation of Section-9.1, a Glasserman type simulation under $\mathbb{P}_0$ would in the domestic economy, have SDEs for the $V(t, T_j)$ for $j = @(t), .., n$ like

$$\frac{dV(t, T_j)}{V(t, T_j)} = \left[\xi^*(t, T_j) - \sum_{k=@(t)}^{j} \phi \left\{ \frac{V(t, T_k)}{Z(t, T_k)} \right\} \xi^*(t, T_k) \right] dW_0(t),$$

$$Z(t, T_k) = \frac{B(t, T_k)}{M(t)} = V(t, T_k) + \Pi_k^k V(t, T_{k+1}) ... + \ \Pi_k^{n-1} V(t, T_n),$$

and in the foreign economy, would have SDEs for the $V^f(t,T_j)$ for $j = @(t),..,n$ like

$$\frac{dV^f(t,T_j)}{V^f(t,T_j)} = \left[\xi^{f*}(t,T_j) - \sum_{k=@(t)}^{j} \phi\left\{\frac{V^f(t,T_k)}{Z^f(t,T_k)}\right\}\xi^{f*}(t,T_k)\right] dW_0^f(t),$$

$$Z^f(t,T_k) = \frac{B^f(t,T_k)}{M^f(t)} = \left\{\begin{array}{c} V^f(t,T_k) + \Pi_k^{f,k} V^f(t,T_{k+1}) + \\ ... + \Pi_k^{f,n-1} V^f(t,T_n) \end{array}\right\},$$

with the foreign BM $W_0^f(t)$ connected to the driving BM $W_0(t)$ under $\mathbb{P}_0$ by (14.12)

$$dW_0^f(t) = dW_0(t) - v_@(t)\,dt.$$

That permits simulation of domestic and foreign rates because the deterministic volatility $v_@(t)$ will be specified for all t through the calibration.

To simulate the FX, proceed as follows. On the current interval $t \in (T_{@-1},\ T_@]$ FX forwards $S_{T_j}(t)$ maturing at $T_j > T_@$ are connected to $S_{T_@}(t)$ through the parity relationship (14.6)

$$S_{T_j}(t) = \frac{F_{T_@}^f(t,T_j)}{F_{T_@}(t,T_j)} S_{T_@}(t) \qquad t \in (T_{@-1},\ T_@],$$

and the domestic and foreign interest rate forward contracts

$$F_{T_@}(t,T_j) = \frac{Z(t,T_j)}{Z(t,T_@)}, \qquad F_{T_@}^f(t,T_j) = \frac{Z^f(t,T_j)}{Z^f(t,T_@)}.$$

The SDE for the immediately maturing FX forward contract $S_{T_@}(t) = S_@(t)$ is

$$\frac{dS_@(t)}{S_@(t)} = \nu_@^*(t)\,dW_0(t), \qquad t \in (T_{@-1},\ T_@],$$

with initial condition at the start of the interval $t = T_{@-1}$, given by the finishing value of the previously maturing contract $S_@(t)$ according to

$$S_{T_@}(T_{@-1}) = \frac{F_{T_{@-1}}^f(T_{@-1},T_@)}{F_{T_{@-1}}(T_{@-1},T_@)} S_{T_{@-1}}(T_{@-1}).$$

And, of course, each contract $S_{T_j}(t)$ *dies* at its maturity T_j.

The overall modus-operandi is therefore to use the same Brownian motion trajectory W_0 to first simulate domestic rates, then the foreign rates, and finally the FX forward contracts, the latter according to the scheme

$S_@(t)$	$T_@ = T_1$	$T_@ = T_2$	$T_@ = T_3$
$S_{T_1}(t)$	$\frac{dS_{T_1}(t)}{S_{T_1}(t)} = \nu^*_{T_1}(t)\,dW_0(t)$	*dead*	*dead*
$S_{T_2}(t)$	$S_{T_2}(t) = \frac{F^f_{T_1}(t,T_2)}{F_{T_1}(t,T_2)} S_{T_1}(t)$	$\frac{dS_{T_2}(t)}{S_{T_2}(t)} = \nu^*_{T_2}(t)\,dW_0(t)$	*dead*
$\vdots$	$\vdots$	$\vdots$	*live*
$S_{T_j}(t)$	$S_{T_j}(t) = \frac{F^f_{T_1}(t,T_j)}{F_{T_1}(t,T_j)} S_{T_1}(t)$	$S_{T_j}(t) = \frac{F^f_{T_2}(t,T_j)}{F_{T_2}(t,T_j)} S_{T_2}(t)$	*live*

$\cdots$	$T_@ = T_{j-1}$	$T_@ = T_j$	$S_@(t)$
$\ddots$	$\vdots$	$\vdots$	$\vdots$
$\vdots$	*dead*	*dead*	$S_{T_{j-2}}(t)$
live	$\frac{dS_{T_{j-1}}(t)}{S_{T_{j-1}}(t)} = \nu^*_{T_{j-1}}(t)\,dW_0(t)$	*dead*	$S_{T_{j-1}}(t)$
live	$S_{T_j}(t) = \frac{F^f_{T_{j-1}}(t,T_j)}{F_{T_{j-1}}(t,T_j)} S_{T_{j-1}}(t)$	$\frac{dS_{T_j}(t)}{S_{T_j}(t)} = \nu^*_{T_j}(t)\,dW_0(t)$	$S_{T_j}(t)$

14.5 Cross-economy correlation

We now extend to two economies, and also include FX, the method of getting forward correlation from swaprate correlation that was developed in Section-6.3.

Assume uniform coverage δ with both foreign and domestic notes coinciding on the floating nodes $T_j = j\delta$ $(j = 0, 1, ..., N-1)$, and as usual add an f superfix to domestic variables to denote the equivalent foreign variable.

Using the relationship (14.12) between the domestic $\mathbb{P}_0$ and foreign $\mathbb{P}^f_0$ spot measures, under the domestic spot measure $\mathbb{P}_0$ relevant SDEs for the domestic and foreign swaprates will have form

$$\frac{d\omega(t,j)}{\omega(t,j)+\alpha} = (drift)\,dt + \sigma^*(j)\,dW_0(t), \tag{14.19}$$
$$\frac{d\omega^f(t,j)}{\omega^f(t,j)+\alpha^f} = (drift)^f\,dt + \sigma^{f*}(j)\,dW_0(t),$$

where α and α^f are constant average estimates of the domestic and foreign swaprate shifts, and the volatilities $\sigma(j)$ and $\sigma^f(j)$ are

$$\sigma(j) = \sum_{k=0}^{j} c_{j,k}\,\xi(x_k), \quad \sigma^f(j) = \sum_{k=0}^{j} c^f_{j,k}\,\xi^f(x_k), \quad j = 0, 1, ..., N-1$$

in which, on the domestic side,

$$\xi(t,T) = \xi(T-t) = \xi(x), \qquad \textit{and}$$
$$c_{j,k} = \begin{cases} \dfrac{B(0,x_{k+1})\,K(0,x_k)}{\sum_{k=0}^{j} B(0,x_{k+1})\,K(0,x_k)} & \textit{for} \quad k = 0,..,j, \\ 0 \quad \textit{for} \quad k = j+1,..,N-1, \end{cases}$$

and, on the foreign side,

$$\xi^f(t,T) = \xi^f(T-t) = \xi^f(x), \qquad \textit{and}$$
$$c^f_{j,k} = \begin{cases} \dfrac{B^f(0,x_{k+1})\,K^f(0,x_k)}{\sum_{k=0}^{j} B^f(0,x_{k+1})\,K^f(0,x_k)} & \textit{for} \quad k = 0,..,j, \\ 0 \quad \textit{for} \quad k = j+1,..,N-1. \end{cases}$$

To condense equations to matrix form below, it will be convenient to introduce the $(2N+1)\times(2N+1)$ upper triangular and non-singular matrix $\mathbf{C}$ constructed from the $c_{j,k}$ and $c^f_{j,k}$ according to

$$C = \begin{pmatrix} c_{0,0} & 0 & \cdots & 0 \\ c_{1,0} & c_{1,1} & \cdots & 0 \\ \vdots & \vdots & \ddots & \vdots \\ c_{N-1,0} & c_{N-1,1} & \cdots & c_{N-1,N-1} \end{pmatrix}$$
$$C^f = \begin{pmatrix} c^f_{0,0} & 0 & \cdots & 0 \\ c^f_{1,0} & c^f_{1,1} & \cdots & 0 \\ \vdots & \vdots & \ddots & \vdots \\ c^f_{N-1,0} & c^f_{N-1,1} & \cdots & c^f_{N-1,N-1} \end{pmatrix},$$
$$\mathbf{C} = \begin{pmatrix} C & 0 & 0 \\ 0 & C^f & 0 \\ 0 & 0 & 1 \end{pmatrix}.$$

From (6.8), the inverses of C and C^f are respectively C^{-1} and $C^{f\,-1}$ as given by (6.10). Hence the inverse of $\mathbf{C}$ is the $(2N+1)\times(2N+1)$ matrix

$$\mathbf{C}^{-1} = \begin{pmatrix} C^{-1} & 0 & 0 \\ 0 & C^{f\,-1} & 0 \\ 0 & 0 & 1 \end{pmatrix}.$$

Our assumption that $\nu_{T_@}(t) = \nu_0(t)$ is deterministic means the relevant data for correlation analysis must be the timeseries of values of the next maturing forward FX contract $S_{T_@}(t) = S_0(t)$. That can be easily constructed from the spot and domestic and foreign three month discount rate timeseries using (14.6), or one can simply confuse the spot data with the forwards because successive differences as used in our quadratic variation estimator below will be about the same. Setting $T_{@} = t+\delta$ and assuming the forward volatility is a constant so as to be homogeneous, that is

$$\nu_{T_{@}}(t) = \nu_0(t) = \nu_0,$$

a formal SDE for the *relative FX forward contract*

$$S_\delta(t) = S_T(t)|_{T=t+\delta}$$

can then be obtained in the usual way

$$\frac{dS_\delta(t)}{S_\delta(t)} = (drift)\,dt + \nu_0^* dW_0(t). \tag{14.20}$$

Using (14.19) and (14.20) quadratic variation estimators for the covariances of domestic rates, foreign rates, domestic against foreign rates, FX against domestic rates, and FX against foreign rates are then respectively:

$$Q = (Q_{j,k}) = \left(\frac{1}{\tau}\int_0^\tau \frac{d\langle\omega(\cdot,j),\omega(\cdot,k)\rangle(t)}{[\omega(t,j)+\alpha][\omega(t,k)+\alpha]}\right),$$

$$Q^f = \left(Q^f_{j,k}\right) = \left(\frac{1}{\tau}\int_0^\tau \frac{d\langle\omega^f(\cdot,j),\omega^f(\cdot,k)\rangle(t)}{[\omega^f(t,j)+\alpha^f][\omega^f(t,k)+\alpha^f]}\right),$$

$$X = (X_{j,k}) = \left(\frac{1}{\tau}\int_0^\tau \frac{d\langle\omega(\cdot,j),\omega^f(\cdot,k)\rangle(t)}{[\omega(t,j)+\alpha][\omega^f(t,k)+\alpha^f]}\right),$$

$$Y = (Y_j) = \left(\frac{1}{\tau}\int_0^\tau \frac{d\langle\omega(\cdot,j),S_\delta(\cdot)\rangle(t)}{[\omega(t,j)+\alpha]\,S_\delta(t)}\right),$$

$$Y^f = \left(Y^f_j\right) = \left(\frac{1}{\tau}\int_0^\tau \frac{d\langle\omega^f(\cdot,j),S_\delta(\cdot)\rangle(t)}{[\omega^f(t,k)+\alpha^f]\,S_\delta(t)}\right),$$

$$Z = \frac{1}{\tau}\int_0^\tau \frac{d\langle S_\delta(\cdot)\rangle(t)}{S^2_\delta(t)}.$$

These can be assembled into the $(2N+1)\times(2N+1)$ *cross-economy covariance* matrix

$$\mathbf{Q} = \begin{pmatrix} Q & X & Y \\ X & Q^f & Y^f \\ Y & Y^f & Z \end{pmatrix} = \mathbf{\Omega}\,\mathbf{\Omega}^T$$

which by principal component analysis can be decomposed in terms of its rank r square-root $\mathbf{\Omega}$ (a $(2N+1)\times r$ matrix). Generally one seems to need $r \cong 7$ to have enough flexibility to jointly cover action in domestic and foreign rates and also the FX. The entries in the first N rows of $\mathbf{\Omega}$ are the domestic swaprate volatilities $\sigma(\cdot)$, the second N entries the foreign swaprate volatilities

$\sigma^f(\cdot)$, and the final entry the constant FX volatility ν_0, and are connected to the domestic and foreign forward volatilities via

$$\mathbf{\Omega} = \begin{pmatrix} \sigma(0) \\ \vdots \\ \sigma(N-1) \\ \sigma^f(0) \\ \vdots \\ \sigma^f(N-1) \\ \nu_0 \end{pmatrix} = \begin{pmatrix} C & 0 & 0 \\ 0 & C^f & 0 \\ 0 & 0 & 1 \end{pmatrix} \begin{pmatrix} \xi(x_0) \\ \vdots \\ \xi(x_{N-1}) \\ \xi^f(x_0) \\ \vdots \\ \xi^f(x_{N-1}) \\ \nu_0 \end{pmatrix} = \mathbf{C\Xi}.$$

In turn, that yields:

1. The corresponding domestic rate $\xi(x)$ and foreign rate $\xi^f(x)$ volatility functions at $x_j = \delta j$, and the constant FX volatility ν_0 via the $(2N+1) \times r$ matrix

$$\begin{pmatrix} \xi(x_0) \\ \vdots \\ \xi(x_{N-1}) \\ \xi^f(x_0) \\ \vdots \\ \xi^f(x_{N-1}) \\ \nu_0 \end{pmatrix} = \mathbf{\Xi} = \mathbf{C}^{-1}\mathbf{\Omega} = \begin{pmatrix} C^{-1} & 0 & 0 \\ 0 & {C^f}^{-1} & 0 \\ 0 & 0 & 1 \end{pmatrix} \begin{pmatrix} \sigma(0) \\ \vdots \\ \sigma(N-1) \\ \sigma^f(0) \\ \vdots \\ \sigma^f(N-1) \\ \nu_0 \end{pmatrix},$$

   from which the vector volatility functions $\xi(x)$ and $\xi^f(x)$ can be interpolated for all x.

2. The corresponding cross-economy $(2N+1) \times (2N+1)$ *implied forward covariance* matrix

$$\mathbf{q} = \mathbf{\Xi\Xi}^T = \mathbf{C}^{-1}\mathbf{\Omega\Omega}^T \left(\mathbf{C}^{-1}\right)^T = \mathbf{C}^{-1}\mathbf{Q}\left(\mathbf{C}^{-1}\right)^T.$$

3. The cross-economy $(2N+1) \times (2N+1)$ *implied forward correlation* matrix R as

$$R = (\operatorname{diag} \mathbf{q})^{-\frac{1}{2}}\ \mathbf{q}\ (\operatorname{diag} \mathbf{q})^{-\frac{1}{2}}. \tag{14.21}$$

4. The domestic $c(\cdot)$ and foreign $c^f(\cdot)$ *correlation functions*

$$c(T-t) = c(x) = \frac{\xi(x)}{\|\xi(x)\|}, \qquad c^f(T-t) = c^f(x) = \frac{\xi^f(x)}{\left\|\xi^f(x)\right\|}.$$

14.6 Pedersen type cross-economy calibration

This section is based on a lecture note by Dun [41] who extended the Pedersen calibration of Section-7.6 to the cross-economy model.

At least three factors (one for domestic rates, one for foreign rates and one for FX) and more usually seven to nine factors are needed in a cross-economy model to properly capture correlations between the forwards in both economies and the FX rates, all of which must be calibrated to interest rate and FX implied volatilities subject to the volatility parity relationship.

This is a complex structure with some data constraints:

1. Reliable implied volatilities for FX forward options out to maturities of 5 to 10 years usually exist, and together with domestic and foreign interest rate implied volatilities constitute a full set of data to which to calibrate.

2. Beyond 5 to 10 year maturities, where only interest rate volatilities are dependable, parity is used to define the FX forward volatilities. But to do that, and produce a successful and believable calibration, we need some notion of how FX forward implied volatilities might behave at more distant maturities (say beyond 10 years).

Using (14.16), the behavior with respect to T of the FX implied volatility $\alpha(T)$ can be gauged with an illustrative example in which we suppose that domestic and foreign initial rates are flat, the shift is flat, all coverages are equal, and domestic, foreign and FX volatilities are flat with each determined by one of three independent factors. That is, for all t and T_j suppose

$$\begin{aligned} \delta_j = \delta_j^f = \delta, \qquad & a(T_j) = a, \quad a^f(T_j) = a^f, \qquad & \nu_{@}(t) = \nu\ (1,0,0), \\ K(0,T_j) = K \quad \Rightarrow \quad & h_j(0) = \frac{\delta(K+a)}{[1+\delta K]} = h \qquad & \xi(t,T_j) = \xi\ (0,1,0), \\ K^f(0,T) = K^f \quad \Rightarrow \quad & h_j^f(0) = \frac{\delta\left(K^f + a^f\right)}{[1+\delta K^f]} = h^f \qquad & \xi^f(t,T) = \xi^f\ (0,0,1). \end{aligned}$$

In that case for $@(t) = T_1$ Section-14.4 yields

$$\begin{aligned} \nu_{T_j}(t) &\cong \nu\ (1,0,0) + (j-1)\,h\xi\ (0,1,0) - (j-1)\,h^f\xi^f\ (0,0,1), \\ \Rightarrow \quad \left|\nu_{T_j}(t)\right|^2 &= \nu^2 + (j-1)^2\left[(h\xi)^2 + \left(h^f\xi^f\right)^2\right], \\ \Rightarrow \quad \alpha^2(T_j) = \frac{1}{j\delta}\int_0^{T_j}\left|\nu_{T_j}(t)\right|^2 dt &= \nu^2 + (j-1)^2\left[(h\xi)^2 + \left(h^f\xi^f\right)^2\right]. \end{aligned}$$

The graph of the implied volatility $\alpha(T_j)$ against $(j-1)$ is a hyperbola, with

$$\lim_{j\to\infty} \alpha(T_j) = (j-1)\sqrt{(h\xi)^2 + \left(h^f\xi^f\right)^2}. \qquad (14.22)$$

Hence forward FX implied volatilities will tend to be concave up and asymptotically linear with maturity, suggesting that for more *distant maturities* where no market data exists, it is reasonable to require that for $T_j > 10$ say,

$$A\,\alpha(T_j) \cong (j-1)$$

where A is some constant that will be determined during calibration. That is, we will let the interest rate volatility data and a linear constraint determine the long term FX implied volatilities.

We now generalize the Pedersen method of Section-7.6 to work in a cross-economy framework. The ingredients are:

1. Pseudo-homogeneous volatility functions for domestic and foreign rates like

$$\xi(t,x) = \xi(t,T)|_{T=t+x} \qquad \textit{and} \qquad \xi^f(t,x) = \xi^f(t,T)\Big|_{T=t+x}$$

   and a deterministic instantaneous forward volatility $\nu_{T_@}(t) = \nu_@(t)$ for the next maturing forward FX contract.

2. The historical correlation matrix R given by (14.21) for correlations between domestic and foreign rates and FX with the irrelevant first row and column stripped so its size is $(2N-1)\times(2N-1)$.

3. Term structures of shift $a(T)$ and $a^f(T)$ chosen to fit skews in the domestic and foreign economies as outlined in Section-7.1.

4. An $(N-1)\times(N-1)$ matrix from which the target domestic caplet and swaption volatilities (targetD_i) are selected, a similar sized $(N-1)\times(N-1)$ matrix for the target foreign caplet and swaption volatilities (targetF_i), and an $(N-1)$ vector from which are selected target FX forward implied volatilities (targetFX_i). We then suppose these three matrices are assembled side-by-side (domestic, foreign, FX) into a $(N-1)\times(2N-1)$ matrix from which the combined targets are selected.

The optimizer in this algorithm will vary the $(N-1)\times(2N-1)$ matrix X comprising the $(N-1)(2N-1)$ parameters,

$$X = (X_{k,\ell}) \qquad k = 1,..,N-1; \quad \ell = 1,..,2N-1$$

and also a further parameter A for the linear constraint on distant FX volatilities. Each parameter $X_{k,\ell}$ in the first (respectively second) $(N-1)\times(N-1)$

block of X is approximately the magnitude $|\xi(t,x)|$ of the domestic (respectively $\left|\xi^f(t,x)\right|$ of the foreign) vector volatility function on certain intervals, while the last column of X is approximately the magnitude $|\nu_@(t)|$ of the FX vector volatility function. Specifically we suppose $\xi(t,x)$, $\xi^f(t,x)$ and $\nu_@(t)$ are piecewise constant with

$$\begin{array}{lll} |\xi(t,x)| \cong X_{k,\ell} & \left|\xi^\ell(t,x)\right| \cong X_{k,\ell+N-1} & |\nu_@(t)| \cong X_{k,2N-1} \\ \textit{for} \quad t \in (T_{k-1}, T_k], & x \in (x_{\ell-1}, x_\ell] & k,\ell = 1,..,N-1. \end{array}$$

The *objective function* $\mathrm{obj}(X,A)$ for the optimizer involves varying the matrix X and the scalar A and has a number of components of form

$$\begin{aligned} &\mathrm{obj}(X,A) = \\ &w_D \ \mathrm{bestfit}_D(X) + w_{D1} \ \mathrm{smooth}_D^{(1)}(X) + w_{D2} \ \mathrm{smooth}_D^{(2)}(X) \\ &+w_F \ \mathrm{bestfit}_F(X) + w_{F1} \ \mathrm{smooth}_F^{(1)}(X) + w_{F2} \ \mathrm{smooth}_F^{(2)}(X) \\ &\qquad +w_{FX} \ \mathrm{bestfit}_{FX}(X) + w_{FX2} \ \mathrm{smooth}_{FX}^{(2)}(X) \\ &\qquad\qquad +w_L \ \mathrm{distant}(A) \end{aligned} \tag{14.23}$$

where the weights w_D, w_{D1}, w_{D2}, w_F, w_{F1}, w_{F2}, w_{FX}, w_{FX2} and w_L are chosen to get a balance between *bestfit* to the target caplets and swaptions, *smoothness* of X in both row and column directions, and linearity of distant FX implied volatilities. The domestic rate components are

$$\begin{aligned} \mathrm{bestfit}_D(X) &= \sum_i \left| \frac{\mathrm{swpnD}_i(X)}{\mathrm{targetD}_i} - 1 \right|^2, \\ \mathrm{smooth}_D^{(1)}(X) &= \sum_{k=1}^{N-1} \sum_{\ell=2}^{N-1} \left| \frac{X_{k,\ell}}{X_{k,\ell-1}} - 1 \right|^2, \\ \mathrm{smooth}_D^{(2)}(X) &= \sum_{k=2}^{N-1} \sum_{\ell=1}^{N-1} \left| \frac{X_{k,\ell}}{X_{k-1,\ell}} - 1 \right|^2, \end{aligned}$$

the foreign rate components are

$$\begin{aligned} \mathrm{bestfit}_F(X) &= \sum_i \left| \frac{\mathrm{swpnF}_i(X)}{\mathrm{targetF}_i} - 1 \right|^2, \\ \mathrm{smooth}_F^{(1)}(X) &= \sum_{k=1}^{N-1} \sum_{\ell=N+1}^{2N-2} \left| \frac{X_{k,\ell}}{X_{k,\ell-1}} - 1 \right|^2, \\ \mathrm{smooth}_F^{(2)}(X) &= \sum_{k=2}^{N-1} \sum_{\ell=N+1}^{2N-2} \left| \frac{X_{k,\ell}}{X_{k-1,\ell}} - 1 \right|^2, \end{aligned}$$

the FX components are

$$\text{bestfit}_{FX}(X) = \sum_i \left| \frac{\text{ivolFX}_i(X)}{\text{targetFX}_i} - 1 \right|^2,$$

$$\text{smooth}^{(2)}_{FX}(X) = \sum_{k=2}^{N-1} \left| \frac{X_{k,2N-1}}{X_{k-1,2N-1}} - 1 \right|^2,$$

and the distant FX component is

$$\text{distant}(A) = \sum_{T_j > 10} \left| A \frac{\alpha(T_j)}{(j-1)} - 1 \right|^2.$$

Note that all components of the objective function are designed to be independent of scale, making the weights w_D, w_{D1}, w_{D2}, w_F, w_{F1}, w_{F2}, w_{FX}, w_{FX2} and w_L comparable and thus easier to adjust to emphasize different aspects of the calibration (like making w_D, w_F and w_{FX} larger to get a better fit at the cost of smoothness of the volatility function).

From X and the historical correlation matrix R, the domestic $\text{swpnD}_i(X)$ and foreign $\text{swpnF}_i(X)$ swaption implied volatilities, and also the FX forward implied volatilities $\text{ivolFX}_i(X)$ are computed by first constructing volatility functions $\xi(\cdot)$, $\xi^f(\cdot)$ and $\nu_{@}(\cdot)$, and then pricing in the standard fashion. For simplicity, fixed and floating side nodes will be assumed to coincide with $T_j = \delta j$ for $j = 0, 1, .., N$ in both foreign and domestic economies.

For each k corresponding to *the* k^{th} *calendar time slice* $t \in (T_{k-1}, T_k]$, for the current X the $(2N-1) \times (2N-1)$ matrix

$$C_k = \text{diag}(X_{k,\ell})\ R\ \text{diag}(X_{k,\ell}) \qquad (\ell = 1, .., 2N-1)$$

(where for an n-vector X, $\text{diag}\, X$ is an $n \times n$ matrix with X along the diagonal and zeros elsewhere) is a covariance. So we can eigenvalue decompose C_k letting $\Gamma^{(1)}_{\cdot}, ..., \Gamma^{(r)}_{\cdot}$ be the first r eigenvectors (ordered by eigenvalue size) multiplied by the square root of the corresponding eigenvalue, so that if

$$\Gamma = \left(\Gamma^{(1)}_\ell, ..., \Gamma^{(r)}_\ell\right), \quad \textit{for} \quad \ell = 1, .., 2N-1$$

$$\textit{then} \qquad \Gamma^T\, \Gamma \cong C_k$$

(the equation is approximate because we have discarded some eigenvectors). For the kth slice of calendar time $t \in (T_{k-1}, T_k]$, now construct the volatility functions $\xi(\cdot)$, $\xi^f(\cdot)$ and $\nu_{@}(\cdot)$ from Γ corresponding to the current value of X by setting for $\ell = 1, .., N-1$ and $x \in (x_{\ell-1}, x_\ell]$

$$\xi(t, x) = \xi_{k.\ell} = \left(\Gamma^{(1)}_\ell, ..., \Gamma^{(r)}_\ell\right)$$

$$\xi^f(t, x) = \xi^f_{k.\ell} = \left(\Gamma^{(1)}_{N-1+\ell}, ..., \Gamma^{(r)}_{N-1+\ell}\right)$$

$$\nu_{@}(t) = \nu_k = \left(\Gamma^{(1)}_{2N-1}, ..., \Gamma^{(r)}_{2N-1}\right)$$

The full volatility function $\xi(t,x) = \xi_{k.\ell}$ covering all calendar times t simply requires this step to be repeated for each slice of calendar time $t \in (T_{k-1}, T_k]$ for $k = 1, 2, ..$

Note that if $t \in (T_{k-1}, T_k]$ the relative maturity x corresponding to T_j is bounded, that is

$$t \in (T_{k-1}, T_k] \quad \textit{and} \quad x + t = T_j \quad \Rightarrow \quad x \in [T_j - T_k,\ T_j - T_{k-1}).$$

That means for each quarterly block of calendar time, the domestic volatilities $\xi(t, T_j)$ and foreign volatilities $\xi^f(t, T_j)$ at absolute maturity T_j are also constant because for $t \in (T_{k-1}, T_k]$

$$\begin{aligned} \xi(t, T_j)\big|_{T_j = t+x} &= \xi(t, x_{j-k+1}) = \xi_{k,\ j-k+1}, \\ \xi^f(t, T_j)\Big|_{T_j = t+x} &= \xi^f(t, x_{j-k+1}) = \xi^f_{k,\ j-k+1}. \end{aligned}$$

Caplet and swaption implied volatilities (or their zetas) in both domestic and foreign economies are then easily computed from (4.16), because for $T \leq T_{j1}, T_{j2}$ they are a linear combination of integrals like

$$\begin{aligned} \int_0^T \xi(t, T_{j1})^* \xi(t, T_{j2})\, dt &= \sum_k \int_{T_{k-1}}^{T_k} \xi(t, T_{j1})^* \xi(t, T_{j2})\, dt \\ &= \delta \sum_k \xi^*_{k,\ j1-k+1}\ \xi_{k,\ j2-k+1}, \\ \int_0^T \xi^f(t, T_{j1})^* \xi^f(t, T_{j2})\, dt &= \sum_k \int_{T_{k-1}}^{T_k} \xi^f(t, T_{j1})^* \xi^f(t, T_{j2})\, dt \\ &= \delta \sum_k \xi^{f\ *}_{k,\ j1-k+1}\ \xi^f_{k,\ j2-k+1}. \end{aligned}$$

Similarly from (14.18) the zeta of the FX forward option maturing at T_j

will be a linear combination of expressions (for $\ell, l1, l2 \geq k$) like

$$\int_{T_{k-1}}^{T_k} |\nu_{T_k}(t)|^2 \, dt = \delta |\nu_k|^2,$$

$$\int_{T_{k-1}}^{T_k} \nu_{T_k}^*(t) \, \xi(t, T_\ell) \, dt = \delta \nu_k^* \xi_{k,\ \ell-k+1},$$

$$\int_{T_{k-1}}^{T_k} \nu_{T_k}^*(t) \, \xi^f(t, T_\ell) \, dt = \delta \nu_k^* \xi_{k,\ \ell-k+1}^f,$$

$$\int_{T_{k-1}}^{T_k} h_{\ell 1}(0) \, h_{\ell 2}(0) \, \xi^*(t, T_{\ell 1}) \, \xi(t, T_{\ell 2}) \, dt$$
$$= \delta h_{\ell 1}(0) \, h_{\ell 2}(0) \, \xi_{k,\ \ell 1-k+1}^* \xi_{k,\ \ell 2-k+1},$$

$$\int_{T_{k-1}}^{T_k} h_{\ell 1}^f(0) \, h_{\ell 2}^f(0) \, \xi^{f\ *}(t, T_{\ell 1}) \, \xi^f(t, T_{\ell 2}) \, dt$$
$$= \delta h_{\ell 1}^f(0) \, h_{\ell 2}^f(0) \, \xi_{k,\ \ell 1-k+1}^{f\ *} \xi_{k,\ \ell 2-k+1}^f,$$

$$\int_{T_{k-1}}^{T_k} h_{\ell 1}(0) \, h_{\ell 2}(0) \, \xi^{f\ *}(t, T_{\ell 1}) \, \xi(t, T_{\ell 2}) \, dt$$
$$= \delta h_{\ell 1}^f(0) \, h_{\ell 2}(0) \, \xi_{k,\ \ell 1-k+1}^{f\ *} \xi_{k,\ \ell 2-k+1}.$$

The overall modus-operandi for Pedersen's algorithm is thus:

Step-1 Start X either with *yesterday's* values, or with the magnitudes of the domestic and foreign historical volatility functions $|\xi(x_f)|$ and $\left|\xi^f(x_f)\right|$, and magnitude of the historical spot FX volatility ν. Start A either with *yesterday's* value or a number estimated from the asymptote (14.22) in the simple example above.

Step-2 From X compute the implied volatilities $\text{swpnD}_i(X)$ and $\text{swpnF}_i(X)$ of the target domestic and foreign caplets and swaptions, and implied volatilities $\text{ivolFX}_i(X)$ of the target FX forward options using the methods described above.

Step-3 Insert those values along with X and A, and the target implied volatilities targetD_i, targetF_i and targetFX_i into the objective function (14.23) and feed into the optimizer to generate new trial values of X and A.

Step-4 Return to Step-2 and iterate until the desired fit and degree of smoothness are obtained. Then extract the final domestic and foreign rate volatilities $\xi(t, x)$ and $\xi^f(t, x)$, and the FX volatility $\nu_@(t)$ as the desired *bestfit* volatility functions.

Chapter 15

Inflation

A major application of the cross-economy BGM model of Chapter-14 is to inflation (see Jarrow et al [68] for the original paper articulating cross-economy HJM to inflation), where the *consumer price index (CPI)* takes the place of the foreign currency exchange rate. Thus in this chapter the variable $S(t)$ denotes the CPI, which is the price in inflatable *dollars* in our nominal-world of one inflation proof *zlotty* in real-world, that is $\mathcal{Z}1 = \$S(t)$. As in the cross-economy model, superfix f to nominal-world variables to denote real-world variables.

Calibration involves stripping nominal and real curves, and identifying forward CPI volatility functions (with correlations) along with a satisfactory forward inflation curve. Once that is done, the pricing and hedging of products is usually straightforward.

15.1 TIPS and the CPI

In the US, basic ingredients available to build an inflation model are:

1. Standard Treasury bonds (nominal variables), which can be stripped using standard techniques.
2. The lagged consumer price index CPI.
3. Inflation indexed bonds called *TIPS (Treasury Inflation Protected Securities)*.
4. Inflation information in the form of futures, inflation forecasts or traders' views that the inflation rate is mean-reverting.

The essential idea behind *TIPS* bonds is that their coupons and final redemption be defined in terms of the CPI index at payoff time. Specifically, if a *TIPS* bond B_{TIPS} is issued at time T_0 when the CPI was $S(T_0)$, then

- A coupon payment of real-world $\mathcal{Z}c$ occurring at time T_j pays off in nominal-world dollars as $\$\frac{S(T_j)}{S(T_0)}c$ dollars.

- Final redemption of real-world $\mathcal{Z}1$ at time T_N is floored, paying off in nominal-world dollars as $\$\max\left[1, \frac{S(T_N)}{S(T_0)}\right]$.

Hence the time t nominal-value of a TIPS bond issued at T_0 ($\leq t$) is

$$B_{TIPS}(t;T_0) = \begin{bmatrix} \sum_{j=1}^{N} \mathbf{E}_0\left\{ \frac{1}{\beta(T_j)} \frac{S(T_j)}{S(T_0)} c \middle| \mathcal{F}_t \right\} \\ +\mathbf{E}_0\left\{ \frac{1}{\beta(T_N)} \frac{S(T_N)}{S(T_0)} \middle| \mathcal{F}_t \right\} + \mathbf{E}_0\left\{ \frac{1}{\beta(T_N)} \left[1 - \frac{S(T_N)}{S(T_0)}\right]^+ \middle| \mathcal{F}_t \right\} \end{bmatrix},$$

$$= \begin{bmatrix} \sum_{j=1}^{N} cB(t,T_j)\,\mathbf{E}_{T_j}\left\{ \frac{S(T_j)}{S(T_0)} \middle| \mathcal{F}_t \right\} \\ +B(t,T_N)\,\mathbf{E}_{T_N}\left\{ \frac{S(T_N)}{S(T_0)} \middle| \mathcal{F}_t \right\} + B(t,T_N)\,\mathbf{E}_{T_N}\left\{ \left[1 - \frac{S(T_N)}{S(T_0)}\right]^+ \middle| \mathcal{F}_t \right\} \end{bmatrix},$$

$$= \begin{bmatrix} \frac{1}{S(T_0)} \left\{ \sum_{j=1}^{N} cB(t,T_j)\,S_{T_j}(t) + B(t,T_N)\,S_{T_N}(t) \right\} \\ +B(t,T_N)\,\mathbf{E}_{T_N}\left\{ \left[1 - \frac{S(T_N)}{S(T_0)}\right]^+ \middle| \mathcal{F}_t \right\} \end{bmatrix},$$

$$= \begin{bmatrix} \dfrac{S(t)}{S(T_0)} \left\{ \sum_{j=1}^{N} cB_f(t,T_j) + B_f(t,T_N) \right\} \\ +B(t,T_N)\,\mathbf{E}_{T_N}\left\{ \left[1 - \frac{S(T_N)}{S(T_0)}\right]^+ \middle| \mathcal{F}_t \right\} \end{bmatrix}.$$

So stripping the *TIPS* bond is complicated by the floor on the CPI index for which we need the forward CPI volatility $\nu_{T_N}(t)$ at time T_N. In practice, the option is often ignored because several years after issue usually the CPI $S(t)$ has increased well beyond its initial value $S(T_0)$ putting the floor well out-of-the-money. Nevertheless, for recently issued TIPS it can be a problem.

The CPI is often expressed in terms of the *zero coupon CPI swap rate* (effectively a quarterly average) defined as that value of κ satisfying

$$(1+\delta\kappa)^N = \frac{S(T)}{S(T_0)} \quad \textit{where} \quad T - T_0 = N\,\delta.$$

As an accumulation index the CPI usually increases and so must have a positive drift $\mu(t)$ under the reference measure $\mathbb{P}$. From the modelling point of view that is not unreasonable: recall that Black-Scholes is perfectly compatible with a steeply drifting stock! In the inflation case, however, the CPI cannot climb in isolation from the bond drifts. We have, using (2.1),

$$B^f(t,T) = \frac{S_T(t)}{S(t)} B(t,T), \qquad \frac{dS(t)}{S(t)} = \mu(t)\,dt + \nu^*(t)\,dW(t),$$

$$\frac{dS_T(t)}{S_T(t)} = -\nu_T(t)\,b(t,T)\,dt + \nu_T^*(t)\,dW(t),$$

$$\frac{dB(t,T)}{B(t,T)} = f(t)\,dt + b^*(t,T)\,dW(t),$$

$$\frac{dB^f(t,T)}{B^f(t,T)} = f^f(t)\,dt + b^{f*}(t,T)\,dW^f(t),$$

where, following the reasoning of Section-2.1, the model for $B^f(t,T)$ is similar to that of $B(t,T)$ under $\mathbb{P}$ in nominal-world, but is now under a different reference measure $\mathbb{P}^f$ in real-world. Having different reference measures in real- and nominal-worlds makes sense, because real-world zlottys constitute a different set of numbers with a different set of statistics to zlottys converted to dollars.

Differentiating both sides of the parity relationship and equating volatility and drift terms in the only way that avoids maturity dependence in the drifts, defines the measure change between $\mathbb{P}$ and $\mathbb{P}^f$, and gives connections

$$dW^f(t) = dW(t) - \nu(t)\,dt, \quad b^f(t,T) = -\nu_T(t) + b(t,T) - \nu(t),$$
$$\mu(t) = f(t) - f^f(t).$$

Because the CPI drift $\mu(t)$ is the difference of the bond drifts $f(t)$ and $f^f(t)$ (clearly an analogue of the Fisher equation), it cannot climb too quickly.

15.2 Dynamics of the forward inflation curve

Products of interest usually revolve around the inflation rate; for example, inflation forwards, inflation futures, inflation to Libor swaps, and options on inflation. So understanding the dynamics of inflation contracts will be important. Let $I(t;\delta)$ be the δ-period *inflation rate* and $I_T(t;\delta)$ the T-maturing *forward inflation* contract on inflation, defined respectively by

$$I(t;\delta) = \frac{S(t)}{S(t-\delta)} - 1 \qquad \textit{and}$$
$$I_T(t;\delta) = \mathbf{E}_T\left\{ I(T;\delta) \middle| \mathcal{F}_t \right\} = \mathbf{E}_T\left\{ \frac{S(T)}{S(T-\delta)} \middle| \mathcal{F}_t \right\} - 1.$$

The forward version of interest rate parity applied to inflation

$$F^f_{T-\delta}(t,T) = \frac{S_T(t)}{S_{T-\delta}(t)} F_{T-\delta}(t,T)$$

says that real rates, nominal rates and the forward CPI are closely connected. The relationship is usually written as the so-called *Fisher equation*

$$1 + \delta K(t,T-\delta) = \left[1 + H_T(t,\delta)\right]\left[1 + \delta K^f(t,T-\delta)\right], \tag{15.1}$$
$$\textit{where} \qquad H_T(t,\delta) = \frac{S_T(t)}{S_{T-\delta}(t)} - 1$$

is called the *pseudo-forward* because its forward CPI components are forward contracts, and at time $T-\delta$ it does equal true forward inflation

$$\begin{aligned} I_T(T-\delta;\delta) &= \mathbf{E}_T\left\{\frac{S_T(T)}{S_{T-\delta}(T-\delta)}\middle|\mathcal{F}_{T-\delta}\right\}-1 \\ &= \frac{S_T(T-\delta)}{S_{T-\delta}(T-\delta)}-1 = H_T(T-\delta,\delta). \end{aligned}$$

In words, the Fisher equation (15.1) says

$$(1+\textit{nominal rate}) = (1+\textit{inflation compensator})\times(1+\textit{real rate}).$$

Though $H_T(t;\delta)$ and $I_T(t;\delta)$ are similar, connecting the two at an arbitrary $t<T-\delta$ requires a *convexity correction.* We have

$$1+I_T(t;\delta) = \mathbf{E}_T\left\{\frac{S_T(T)}{S_{T-\delta}(T-\delta)}\middle|\mathcal{F}_t\right\} = \mathbf{E}_T\left\{\frac{S_T(T-\delta)}{S_{T-\delta}(T-\delta)}\middle|\mathcal{F}_t\right\},$$

$$= \frac{S_T(t)}{S_{T-\delta}(t)}\mathbf{E}_T\left\{\frac{\mathcal{E}\left\{\int_t^{T-\delta}\nu_T^*(s)\,dW_T(s)\right\}}{\mathcal{E}\left\{\int_t^{T-\delta}\nu_{T-\delta}^*(s)\,dW_{T-\delta}(s)\right\}}\middle|\mathcal{F}_t\right\} = \frac{S_T(t)}{S_{T-\delta}(t)}\exp\alpha(t,T),$$

where

$$\begin{aligned} \exp\alpha(t,T) &= \mathbf{E}_T\left\{\begin{array}{c}\exp\left\{\int_t^{T-\delta}\nu_{T-\delta}^*(s)\,b(s,T-\delta,T)\,ds\right\} \\ \times\dfrac{\mathcal{E}\left\{\int_t^{T-\delta}\nu_T^*(s)\,dW_T(s)\right\}}{\mathcal{E}\left\{\int_t^{T-\delta}\nu_{T-\delta}^*(s)\,dW_T(s)\right\}}\end{array}\middle|\mathcal{F}_t\right\}, \\ &= \mathbf{E}_T\left\{\begin{array}{c}\exp\left\{\int_t^{T-\delta}\nu_{T-\delta}^*(s)\,b^f(s,T-\delta,T)\,ds\right\} \\ \times\mathcal{E}\left\{\int_t^{T-\delta}\left[\nu_T(s)-\nu_{T-\delta}(s)\right]^*dW_T(s)\right\}\end{array}\middle|\mathcal{F}_t\right\}, \end{aligned}$$

yields the $\mathcal{F}_t$-measurable stochastic convexity variable $\alpha(t,T)$. If forward CPI volatilities $\nu_T(t)$ are all assumed deterministic, and the bond volatility difference $b(s,T-\delta,T)$ is approximated deterministically by setting its stochastic components to their time t values, an approximation to $\alpha(\cdot)$ will be

$$\begin{aligned} \alpha(t,T) &\cong \int_t^{T-\delta}\left[b(s,T-\delta,T)-\nu_T(s)+\nu_{T-\delta}(s)\right]^*\nu_{T-\delta}(s)\,ds, \\ &\cong \left[\tfrac{\delta H(t,T-\delta)}{1+\delta K(t,T-\delta)}\xi^*(T-\delta)\,\nu_{T-\delta}-\nu_T^*\nu_{T-\delta}+\left|\nu_{T-\delta}\right|^2\right](T-\delta-t). \end{aligned}$$

The Fisher equation in terms of the true forward and convexity adjustment is

$$1+\delta K(t,T-\delta) = \exp\{-\alpha(t,T)\}\left[1+I_T(t,\delta)\right]\left[1+\delta K^f(t,T-\delta)\right].$$

REMARK 15.1 In HJM all volatilities are deterministic, so

$$\alpha(0,T)=\int_0^{T-\delta}\left[b(s,T-\delta,T)-\nu_T(s)+\nu_{T-\delta}(s)\right]^*\nu_{T-\delta}(s)\,ds$$

exactly. To roughly estimate the magnitude of $\alpha(\cdot)$ assume the forward CPI volatilities $\nu_T(t)$ are 2%, cash forwards $K(t,T)$ are 6%, the BGM volatility $\xi(t,T)$ is flat at 10%, and $\delta=.25$. Then

$$b(t,T-\delta,T)=\int_{T-\delta}^{T}\sigma(t,u)\,du=\frac{\delta K(t,T)}{1+\delta K(t,T)}\xi(t,T)\cong .25\times .06\times .1,$$
$$\alpha(0,T)\cong(T-\delta)\times .25\times .06\times .1\times .02=.00003\times(T-\delta),$$

giving corrections like $\alpha(0,10)\cong 3$ bpts, $\alpha(0,50)\cong 15$ bpts. □

The dynamics of forward inflation $I_T(t;\delta)$ will be determined by

$$I_T(t;\delta)=\frac{S_T(t)}{S_{T-\delta}(t)}\exp\{\alpha(t,T-\delta,T)\}-1,$$

and is a $\mathbb{P}_T$-martingale. So if we differentiate it and express the resultant SDE in terms of $W_T(t)$, all drift terms must cancel. Hence

$$dI_T(t;\delta)=\left[1+I_T(0;\delta)\right]\left[\nu_T(t)-\nu_{T-\delta}(t)+\nu_\alpha(t)\right]^*dW_T(t),$$

where stochastic $\nu_\alpha(t)$ comes from $\alpha(\cdot)$ and is assumed small. Thus $I_T(t;\delta)$ is approximately a shifted lognormal process with a shift of 1, making it almost normal with approximate small volatility $[\nu_T(t)-\nu_{T-\delta}(t)]$.

REMARK 15.2 That $I_T(t;\delta)$ is normally distributed with low variance, may account for traders' observations that inflation is mean reverting. Moreover, because the forward volatility expression $[\nu_T(t)-\nu_{T-\delta}(t)]$ will affect the value of inflation options, we will need to be careful in our calibration. □

15.2.1 Futures contracts

Similarly to the definition of the forward contract, the *futures contract* $J_T(t,\delta)$ maturing at T on the inflation rate $I(t,\delta)$ is

$$J_T(t,\delta)=\mathbf{E}_0\left\{I(T,\delta)|\,\mathcal{F}_t\right\}=\mathbf{E}_0\left\{\left.\frac{S(T)}{S(T-\delta)}\right|\mathcal{F}_t\right\}-1,$$

so that if, as with the forwards, the CPI volatilities $\nu_T(t)$ are deterministic

$$1+J_T(t,\delta)=\mathbf{E}_0\left\{\left.\frac{S_T(T)}{S_{T-\delta}(T-\delta)}\right|\mathcal{F}_t\right\},$$

$$=\frac{S_T(t)}{S_{T-\delta}(t)}\mathbf{E}_0\left\{\left.\frac{\mathcal{E}\left\{\int_t^T\nu_T^*(s)\,dW_T(s)\right\}}{\mathcal{E}\left\{\int_t^{T-\delta}\nu_{T-\delta}^*(s)\,dW_{T-\delta}(s)\right\}}\right|\mathcal{F}_t\right\}=\frac{S_T(t)}{S_{T-\delta}(t)}\exp\beta(t,T)$$

where

$$\exp\beta(t,T) = \mathbf{E}_0\left\{\left.\frac{\mathcal{E}\left\{\int_t^T \nu_T^*(s)\left[dW_0(s)+b(s,s,T)\ ds\right]\right\}}{\mathcal{E}\left\{\int_t^{T-\delta} \nu_{T-\delta}^*(s)\left[dW_0(s)+b(s,s,T-\delta)\ ds\right]\right\}}\right|\mathcal{F}_t\right\},$$

$$= \mathbf{E}_0\left\{\left.\begin{array}{c}\exp\left[\begin{array}{c}\int_t^T \nu_T^*(s)\,b(s,s,T)\ ds\\ -\int_t^{T-\delta} \nu_{T-\delta}^*(s)\,b(s,s,T-\delta)\ ds\end{array}\right]\\ \times\dfrac{\mathcal{E}\left\{\int_t^{T-\delta}\nu_T^*(s)\,dW_0(s)\right\}}{\mathcal{E}\left\{\int_t^{T-\delta}\nu_{T-\delta}^*(s)\,dW_0(s)\right\}}\end{array}\right|\mathcal{F}_t\right\},$$

$$= \mathbf{E}_0\left\{\left.\begin{array}{c}\exp\left[\begin{array}{c}\int_t^T \nu_T^*(s)\,b(s,s,T)\ ds\\ -\int_t^{T-\delta} \nu_{T-\delta}^*(s)\,b(s,s,T-\delta)\ ds\\ -\int_t^{T-\delta}\left[\nu_T(s)-\nu_{T-\delta}(s)\right]^*\nu_{T-\delta}(s)\,ds\end{array}\right]\\ \times\mathcal{E}\left\{\int_t^{T-\delta}\left[\nu_T(s)-\nu_{T-\delta}(s)\right]^* dW_0(s)\right\}\end{array}\right|\mathcal{F}_t\right\},$$

yields the $\mathcal{F}_0$-measurable stochastic convexity variable $\beta(t,T)$. Approximating the drift deterministically, yields a convexity adjustment like

$$\begin{aligned}&\beta(t,T)\\ &= \left[\begin{array}{c}\int_t^T \nu_T^*(s)\,b(s,s,T)\ ds - \int_t^{T-\delta}\nu_{T-\delta}^*(s)\,b(s,s,T-\delta)\ ds\\ -\int_t^{T-\delta}\left[\nu_T(s)-\nu_{T-\delta}(s)\right]^*\nu_{T-\delta}(s)\,ds\end{array}\right],\\ &= \alpha(t,T)+\int_t^T \nu_T^*(s)\,b(s,s,T)\ ds - \int_t^{T-\delta}\nu_{T-\delta}^*(s)\,b(s,s,T)\,ds.\end{aligned}$$

Thus the Fisher equation expressed in terms of $J_T(t,\delta)$ is

$$1+\delta K(t,T-\delta) = \exp\{-\beta(t,T)\}\left[1+J_T(t,\delta)\right]\left[1+\delta K^f(t,T-\delta)\right].$$

15.2.2 The CME futures contract

With an inflation period of 3 months ($\delta = .25$) the time T settlement futures price of this contract is specified as

$$\begin{aligned}\mathrm{CME}(T) &= 100 - 4\times 100\times\frac{\mathrm{CPI}(T)-\mathrm{CPI}(T-3\,\mathrm{months})}{\mathrm{CPI}(T-3\,\mathrm{months})}\%,\\ &= 100\left(1-4\left[\frac{S(T)}{S(T-\delta)}-1\right]\right) = 100\left(1-4J_T(T,\delta)\right)\%,\\ \Rightarrow\quad \mathrm{CME}(t) &= \mathbf{E}_0\left\{100\left(1-4J_T(T,\delta)\right)\middle|\mathcal{F}_t\right\} = 100\left(1-4J_T(t,\delta)\right)\%.\end{aligned}$$

If $\widetilde{\mathrm{CME}}(t)$ is an estimate of the futures contract using the pseudo-forward

$$\widetilde{\mathrm{CME}}(t) = 100\left(1-4H_T(t,\delta)\right)\%,$$

then the convexity correction required to get a more accurate price is

$$\begin{aligned}\mathrm{CME}(t) - \widetilde{\mathrm{CME}}(t) &= -400\left[J_T(t,\delta) - H_T(t,\delta)\right]\%,\\ &= -400\left[\exp\{\beta(t,T-\delta,T)\} - 1\right]\frac{S_T(t)}{S_{T-\delta}(t)}\%.\end{aligned}$$

Chapter 16

Stochastic Volatility BGM

Our aim in this chapter is not to try to fit a volatility smile exactly, but to add a measure of *convexity* to an existing skew. We develop a shifted-stochastic volatility model with zero correlation between the forwards and their stochastic volatility, in which the shift (rather than the absent correlation) is used to fit the skew.

Independence of the yieldcurve and stochastic volatility drivers simplifies the mathematics, and yields a model that when compared to Shifted BGM:

1. Has only a few more parameters and is only marginally more complex to calibrate using techniques already developed.
2. Has a similar term structure of skew, but no term structure of convexity.
3. Makes forward measure changes in an equally straightforward way.
4. Also permits caplets and swaptions to be priced by a single technique.
5. Uses almost identical methods to compute deltas and vegas.

An added measure of flexibility is that the stochastic volatility or *variance* process can be any reasonable positive process such as the square-root process considered here (other possibilities include, for example, the exponential Ornstein Uhlenbeck processes).

16.1 Construction

Return to (2.1) and suppose that under the reference measure $\mathbb{P}$, all bonds have an additional stochastic component in their volatility depending on a positive *variance process* $V(t)$ like

$$\frac{dB(t,T)}{B(t,T)} = f(t)\,dt + \sqrt{V(t)}b^*(t,T)\,dW(t) \quad t \le T, \tag{16.1}$$

where the multi-dimensional BM $W(t)$ appears only in the bond SDEs, and is independent of the extra one-dimensional BM $U(t)$ driving $V(t)$

$$dV(t) = \mu(t,V(t))\,dt + \gamma(t,V(t))\,dU(t), \quad V(t) > 0 \quad V(0) = 1.$$

Note that $V(0)$ can always be set at 1, otherwise simply re-scale $V(\cdot)$ and $b(\cdot)$. For any zero coupon bond $B(t,S)$ with an earlier maturity $t \le S < T$

$$d\left(\tfrac{B(t,S)}{B(t,T)}\right) \Big/ \left(\tfrac{B(t,S)}{B(t,T)}\right)$$
$$= \sqrt{V(t)}\,[b(t,S) - b(t,T)]^* \left[dW(t) - \sqrt{V(t)}b(t,T)\,dt\right],$$

so we can define a *change of measure* from $\mathbb{P}$ to the *forward measure* $\mathbb{P}_T$ induced by $B(t,T)$ as numeraire by

$$dW_T(t) = dW(t) - \sqrt{V(t)}b(t,T)\,dt \qquad dU_T(t) = dU(t).$$

Note that if $W(t)$ and $U(t)$ were correlated, there would be a non-zero component of the bond volatility $b(\cdot)$ corresponding to $U(t)$, and consequently the BM $U(t)$ driving $V(t)$ would not be the same under both reference and forward measures. Thus bonds discounted by $B(t,T)$ as numeraire are martingales under $\mathbb{P}_T$

$$d\left(\tfrac{B(t,S)}{B(t,T)}\right) \Big/ \left(\tfrac{B(t,S)}{B(t,T)}\right) = \sqrt{V(t)}b(t,S,T)^*\,dW_T(t).$$

For $T < S = T_1$ the *measure change* between $\mathbb{P}_T$ and $\mathbb{P}_{T_1}$ is given by

$$dW_{T_1}(t) = dW_T(t) + \sqrt{V(t)}b(t,T,T_1)\,dt,$$

and the ratio of the bonds is the forward contract or its reciprocal

$$F_T(t,T_1) = \frac{B(t,T_1)}{B(t,T)}, \qquad \frac{1}{F_T(t,T_1)} = \frac{B(t,T)}{B(t,T_1)},$$

which for $t \le T$ have respective SDEs

$$\tfrac{dF_T(t,T_1)}{F_T(t,T_1)} = -\sqrt{V(t)}b(t,T,T_1)^*\,dW_T(t),$$
$$d\left(\tfrac{1}{F_T(t,T_1)}\right) \Big/ \left(\tfrac{1}{F_T(t,T_1)}\right) = \sqrt{V(t)}b(t,T,T_1)^*\,dW_{T_1}(t).$$

We are now in a position to describe a *backward construction* for our model similar to that of Section 3.2. Temporarily simplify notation by indexing variables by maturity and dropping time t, for example

$$B_T = B(t,T) \quad W_T = W_T(t) \quad b_T = b(t,T) \quad a(T) = a_T \quad etc.$$

For $0 \le t \le R < S < T < U$ we will illustrate the method by moving back through the intervals $(T,U]$, $(S,T]$ and $(R,S]$ successively defining the bond volatility differences $[b_T - b_U]$, $[b_S - b_T]$ and $[b_R - b_S]$ in the process. Our *basic assumption* and starting point is that the deterministic volatility functions ξ_R, ξ_S and ξ_T are specified exogenously and are bounded, and that K_T is a martingale under $\mathbb{P}_U$. Thus on the three intervals respectively

$$\delta_T = |(T,U]| \quad \delta_T K_T = \frac{B_T}{B_U} - 1 \quad H_T = K_T + a_T$$
$$dH_T = \sqrt{V} H_T \xi_T^* dW_U,$$
$$\delta_S = |(S,T]| \quad \delta_S K_S = \frac{B_S}{B_T} - 1 \quad H_S = K_S + a_S$$
$$dH_S = H_S \left[\mu_S dt + \sqrt{V}\xi_S^* dW_U\right],$$
$$\delta_R = |(R,S]| \quad \delta_R K_R = \frac{B_R}{B_S} - 1 \quad H_R = K_R + a_R$$
$$dH_R = H_R \left[\mu_R dt + \sqrt{V}\xi_R^* dW_U\right].$$

On the *first interval* $(T,U]$, comparison of volatility terms in

$$\delta_T dH_T = \delta_T \sqrt{V} H_T \xi_T^* dW_U = \delta_T dK_T$$
$$= d\left(\frac{B_T}{B_U}\right) = \left(\frac{B_T}{B_U}\right)\sqrt{V}\left[b_T - b_U\right]^* dW_U$$

identifies the difference in the bond volatilities

$$\left[b_T - b_U\right] = \frac{\delta_T H_T}{1+\delta_T K_T}\xi_T,$$

which must be bounded (assuming $\delta_T a_T < 1$) because the SDE for H_T has a positive solution, and that defines a change of measure from $\mathbb{P}_U$ to $\mathbb{P}_T$ by

$$dW_T = dW_U - \sqrt{V}\left[b_T - b_U\right]dt = dW_U - \frac{\delta_T H_T}{1+\delta_T K_T}\sqrt{V}\xi_T dt.$$

On the *second interval* $(S,T]$, in the equation

$$\delta_S dH_S = \delta_S H_S \left[\mu_S dt + \sqrt{V}\xi_S^* dW_U\right]$$
$$= \delta_S dK_S = d\left(\frac{B_S}{B_T}\right) = d\left(\frac{B_S}{B_U}\Big/\frac{B_T}{B_U}\right)$$
$$= \left(\frac{B_S}{B_T}\right)\sqrt{V}\left[b_S - b_T\right]^*\left\{dW_U - \sqrt{V}\left[b_T - b_U\right]dt\right\},$$

comparison of volatility and drift terms yields

$$\delta_S H_S \xi_S = \left(\frac{B_S}{B_T}\right)\left[b_S - b_T\right], \qquad \textit{and}$$
$$\delta_S H_S \mu_S = -\left(\frac{B_S}{B_T}\right) V \left[b_S - b_T\right]^*\left[b_T - b_U\right] \qquad \Rightarrow$$
$$\mu_S = -V\xi_S^*\left[b_T - b_U\right] \qquad \Rightarrow$$
$$dH_S = H_S\sqrt{V}\xi_S^*\left[dW_U - \sqrt{V}\left[b_T - b_U\right]dt\right] = H_S\sqrt{V}\xi_S^* dW_T.$$

The SDE for H_S has a solution that is strictly positive, which in turn defines the bond volatility difference

$$[b_S - b_T] = \frac{\delta_S H_S}{1+\delta_S K_S}\xi_S,$$

and a change of measure from $\mathbb{P}_U$ to $\mathbb{P}_S$ by

$$\begin{aligned} dW_S &= dW_U - \sqrt{V}\,[b_S - b_U]\,dt = dW_U - \left[\sqrt{V}\,(b_S - b_T) + \sqrt{V}\,(b_T - b_U)\right] dt \\ &= dW_U - \left\{\frac{\delta_S H_S}{1+\delta_S K_S}\sqrt{V}\xi_S + \frac{\delta_T H_T}{1+\delta_T K_T}\sqrt{V}\xi_T\right\} dt. \end{aligned}$$

On the *third interval* $(R, S]$ proceed as on the second.

$$\begin{aligned} \delta_R dH_R &= \delta_R H_R \left[\mu_R dt + \sqrt{V}\xi_R^* dW_U\right] \\ &= \left(\frac{B_R}{B_S}\right)\sqrt{V}\,[b_R - b_S]^* \left\{dW_U - \sqrt{V}\,[b_S - b_U]\,dt\right\} \\ \Rightarrow \qquad dH_R &= H_R\sqrt{V}\xi_R^*\left[dW_U - \sqrt{V}\,[b_S - b_U]\,dt\right] = \sqrt{V}H_R\xi_R^* dW_S \\ \Rightarrow \qquad [b_R - b_S] &= \frac{\delta_R H_R}{1+\delta_R K_R}\xi_R. \end{aligned}$$

As before, the SDE for H_R has a strictly positive solution which defines $[b_R - b_S]$ and the next measure change from $\mathbb{P}_U$ to $\mathbb{P}_R$ by

$$\begin{aligned} dW_R &= dW_U - \sqrt{V}\,[b_R - b_U]\,dt, \\ &= dW_U - \left\{\frac{\delta_S H_S}{1+\delta_S K_S}\sqrt{V}\xi_S + \frac{\delta_T H_T}{1+\delta_T K_T}\sqrt{V}\xi_T + \frac{\delta_R H_R}{1+\delta_R K_R}\sqrt{V}\xi_R\right\} dt. \end{aligned}$$

The measures $\mathbb{P}_R$, $\mathbb{P}_S$ and $\mathbb{P}_T$ thus defined are clearly the forward measures at R, S and T. For example, for any bond B_Q $(Q < T)$ the ratio $\left(\frac{B_Q}{B_T}\right)$ must be a $\mathbb{P}_T$-martingale because

$$d\left(\frac{B_Q}{B_T}\right) = d\left(\frac{B_Q}{B_U}\Big/\frac{B_T}{B_U}\right) = \left(\frac{B_Q}{B_T}\right)\sqrt{V}\,[b_Q - b_T]^*\,dW_T.$$

In general, starting with the *terminal measure* $\mathbb{P}_n$ at the terminal node T_n and working backwards in the above fashion *constructs* a system of equations with each forward $H(t, T_j)$ an exponential martingale under the corresponding forward measure $\mathbb{P}_{j+1}$.

The general structure of the model is now becoming clear; whenever a volatility function occurs in shifted BGM simply multiply it by $\sqrt{V}$ to get the stochastic BGM equivalent. The next section shows the same result holds for our approximations to swaprate volatilities; all the results of Chapter-4 hold in stochastic volatility BGM if volatilities are multiplied by $\sqrt{V_t}$.

16.2 Swaprate dynamics

Recall the swaprate $\omega(t)$ in Chapter-4 was split into *stochastic* $\omega 1(t)$ and *shift* $\omega 2(t)$ parts:

$$\omega(t) = \omega 1(t) - \omega 2(t) \qquad \textit{where}$$

$$\omega 1(t) = \frac{\sum_{j=0}^{N-1} \delta_j F_T(t, T_{j+1}) H(t, T_j)}{\sum_{i=0}^{M-1} \overline{\delta}_i F_T\left(t, \overline{T}_{i+1}\right)} = \sum_{j=0}^{N-1} u_j(t) H(t, T_j) = \sum_{j=0}^{N-1} v_j(t),$$

$$\omega 2(t) = \frac{\sum_{j=0}^{N-1} \delta_j F_T(t, T_{j+1}) \left[a(T_j) - \mu_j\right]}{\sum_{i=0}^{M-1} \overline{\delta}_i F_T\left(t, \overline{T}_{i+1}\right)} = \sum_{j=0}^{N-1} u_j(t) \left[a(T_j) - \mu_j\right].$$

If $\widetilde{\mathbb{P}}_T$ is now the *swaprate measure* equivalent to $\mathbb{P}_T$ induced by $\widetilde{W}_T(t)$, where

$$d\widetilde{W}_T(t) = dW_T(t) + \sqrt{V(t)} \sum_{i=0}^{M-1} \overline{u}_i(t)\, b\left(t, T, \overline{T}_{i+1}\right) dt,$$

then the weights $u_j(t)$ and $\overline{u}_i(t)$ will be low variance $\widetilde{\mathbb{P}}_T$-martingales

$$\frac{du_j(t)}{u_j(t)} = \sqrt{V(t)} \left[-b(t, T, T_{j+1}) + \sum_{i=0}^{M-1} \overline{u}_i(t)\, b\left(t, T, \overline{T}_{i+1}\right)\right]^* d\widetilde{W}_T(t),$$

$$\frac{d\overline{u}_i(t)}{\overline{u}_i(t)} = \sqrt{V(t)} \left[-b\left(t, T, \overline{T}_{i+1}\right) + \sum_{i=0}^{M-1} \overline{u}_i(t)\, b\left(t, T, \overline{T}_{i+1}\right)\right]^* d\widetilde{W}_T(t).$$

Hence the approximation (which is the same as before) for the *shift*

$$\begin{aligned}\omega 2(t) \cong \omega 2(0) &= \frac{\sum_{j=0}^{N-1} \delta_j F_T(0, T_{j+1}) \left[a(T_j) - \mu_j\right]}{\sum_{i=0}^{M-1} \overline{\delta}_i F_T\left(0, \overline{T}_{i+1}\right)} \\ &= \sum_{j=0}^{N-1} u_j(0) \left[a(T_j) - \mu_j\right] = \alpha(0).\end{aligned} \tag{16.2}$$

In a similar fashion, the $v_j(t)$ will also be $\widetilde{\mathbb{P}}_T$-martingales

$$\frac{dv_j(t)}{v_j(t)} = \sqrt{V(t)} \left[\begin{array}{c} \xi(t, T_j) - b(t, T, T_{j+1}) \\ + \sum_{i=0}^{M-1} \overline{u}_i(t)\, b\left(t, T, \overline{T}_{i+1}\right) \end{array}\right]^* d\widetilde{W}_T(t),$$

for which a more stable SDE can be obtained using $u_j(t) \cong u_j(0)$ like

$$\begin{aligned} v_j(t) &= u_j(t) H(t,T_j) = u_j(t)\left[K(t,T_j) + a(T_j)\right] \\ &\cong u_j(t) K(t,T_j) + u_j(0) a(T_j) \\ \Rightarrow \quad \frac{dv_j(t)}{v_j(t)} &\cong \frac{d\left[u_j(t) K(t,T_j)\right]}{u_j(t) K(t,T_j)} \frac{K(t,T_j)}{H(t,T_j)} \end{aligned}$$

$$= \sqrt{V(t)}\left[\xi(t,T_j) + \frac{K(t,T_j)}{H(t,T_j)}\begin{pmatrix} -b(t,T,T_{j+1}) \\ +\sum_{i=0}^{M-1} \overline{u}_i(t)\, b\left(t,T,\overline{T}_{i+1}\right)\end{pmatrix}\right]^* d\widetilde{W}_T(t).$$

Hence the following SDE for the *stochastic part* $\omega 1(t)$ of the swaprate

$$\frac{d\omega 1(t)}{\omega 1(t)} = \sigma(t)\, d\widetilde{W}_T(t), \quad \sigma(t) = \sqrt{V(t)} \sum_{j=0}^{N-1} A_j(t)\, \xi(t,T_j),$$

$$A_j(t) = \left\{ w_j(t) - h_j(t) \sum_{\ell=j}^{N-1} \begin{pmatrix} w_\ell(t) \dfrac{K(t,T_\ell)}{H(t,T_\ell)} \\ -\vec{u}_\ell(t) \sum_{k=0}^{N-1} \frac{K(t,T_k)}{H(t,T_k)} w_k(t) \end{pmatrix} \right\}$$

$$\textit{and} \qquad w_j(t) = \frac{\delta_j F_T(t,T_{j+1}) H(t,T_j)}{\sum_{j=0}^{N-1} \delta_j F_T(t,T_{j+1}) H(t,T_j)}.$$

SDEs for the weights $w_j(t)$ show they are low variance martingales well approximated by their initial values $w_j(0)$, giving the following approximate SDE for the *stochastic part* $\omega 1(t)$ of the swaprate $\omega(t)$

$$\frac{d\omega 1(t)}{\omega 1(t)} = \sqrt{V(t)} \left\{ \sum_{j=0}^{N-1} A_j \xi^*(t,T_j) \right\} d\widetilde{W}_T(t), \tag{16.3}$$

$$A_j = \left\{ w_j - h_j \sum_{\ell=j}^{N-1} \left(w_\ell \frac{K_\ell}{H_\ell} - \vec{u}_\ell \lambda \right) \right\}, \qquad \lambda = \sum_{\ell=0}^{N-1} \frac{K_\ell}{H_\ell} w_\ell$$

$$w_j = w_j(0) \quad h_j = h_j(0) \quad K_\ell = K(0,T_\ell) \quad H_\ell = H(0,T_\ell) \quad \vec{u}_\ell = \vec{u}_\ell(0).$$

Hence in stochastic volatility BGM, caplets and swaptions can be priced by the same method, because given a variance process $V(t)$ satisfying (16.1) the time t value $\text{Cpl}(t)$ of the time T maturing *caplet* struck at κ and paying at T_1 is

$$\text{Cpl}(t) = \delta B(t,T_1)\, \mathbf{E}_{T_1}\left\{ \left[K(T,T) - \kappa\right]^+ \middle| \mathcal{F}_t \right\}, \qquad \textit{where} \tag{16.4}$$

$$H(t,T) = K(t,T) + a(T), \qquad \frac{dH(t,T)}{H(t,T)} = \sqrt{V(t)}\xi^*(t,T)\, dW_{T_1}(t),$$

while from (16.2) and (16.3) the time t value of a *swaption* maturing at T and struck at κ will be

$$\text{pSwpn}(t) = \left\{ \sum_{i=0}^{M-1} \overline{\delta}_i B\left(t, \overline{T}_{i+1}\right) \right\} \widetilde{\mathbf{E}}_T \left\{ \left[\omega(T) - \kappa\right]^+ \middle| \mathcal{F}_t \right\}, \tag{16.5}$$

$$\omega 1(t) \cong \omega(t) + \alpha(0), \qquad \frac{d\omega 1(t)}{\omega 1(t)} = \sqrt{V_t} \left\{ \sum_{j=0}^{N-1} A_j \xi^*(t, T_j) \right\} d\widetilde{W}_T(t).$$

16.3 Shifted Heston options

The solution to the next problem gives prices of caplets and swaptions in stochastic volatility BGM by inserting into (16.6)

$$\xi(t) = |\xi(t,T)| \quad \textit{for caplets and} \quad \xi(t) = \left| \sum_{j=0}^{N-1} A_j \xi^*(t, T_j) \right| \quad \textit{for swaptions}$$

Problem 1 *(The shifted Heston option problem) Find*

$$\mathbf{H}(t,\kappa) = \mathbf{E}\left\{ \left[K(T) - \kappa\right]^+ \middle| \mathcal{F}_t \right\}, \tag{16.6}$$

$$\textit{when} \quad \frac{dK(t)}{K(t) + a} = \sqrt{V(t)}\xi(t)\, dW(t),$$

$$\textit{and} \quad dV(t) = \lambda\left[\mu - V(t)\right] dt + \gamma\sqrt{V(t)} dU(t),$$

$$\textit{with} \quad \lambda, \mu, \gamma > 0 \quad V(0) = 1.$$

Setting $S(t) = \ln(K(t) + a)$ yields the 2-dimensional *affine system*

$$dS(t) = -\frac{1}{2}V(t)\ \xi^2(t)\, dt + \sqrt{V(t)}\xi(t)\, dW(t),$$

$$dV(t) = \lambda\left[\mu - V(t)\right] dt + \gamma\sqrt{V(t)} dU(t), \quad V(0) = 1,$$

which is Markov in terms of the state vector $X(t) = (S(t),\ V(t))^*$.

16.3.1 Characteristic function

Let $\widehat{p}(\theta; t)$, where θ is a complex constant, be the characteristic function of $S(T)$ conditional on $X(t)$

$$\widehat{p}(\theta; t) = \mathbf{E}\left\{\exp(i\theta S(T)) \mid X(t)\right\} \quad \textit{with} \quad \widehat{p}(\theta; T) = \exp(i\theta S(T)).$$

REMARK 16.1 To be totally consistent with our Fourier transform definitions (A.10) we should perhaps write $2\pi\theta$ rather than θ, but that is not how characteristic functions are usually defined. Thus a little care is required in applying $\widehat{p}$ below. □

Because the system is Markov, $\widehat{p}(\cdot)$ must satisfy the backward PDE

$$\widehat{p}_t - \frac{1}{2}\xi^2(t)\,V(t)\,\widehat{p}_s + \lambda\left[\mu - V(t)\right]\,\widehat{p}_v + \frac{1}{2}\xi^2(t)\,V(t)\,\widehat{p}_{ss} + \frac{1}{2}\gamma^2 V(t)\,\widehat{p}_{vv} = 0.$$

and, because the system is affine, $\widehat{p}(\cdot)$ must also have the *affine form*

$$\begin{aligned}\widehat{p}(\theta;t) &= \exp\left(A(t) + B(t)\,S(t) + C(t)\,V(t)\right),\\ A(T) &= C(T) = 0, \quad B(T) = i\theta,\end{aligned}$$

where A, B, C are functions of t (and θ) only. Substituting $\widehat{p}(\cdot)$ into the PDE and setting the coefficients of $S(t)$, $V(t)$ and time dependent terms zero, gives

$$\begin{aligned}&B_t\; S(t) + \left[C_t - \lambda C + \frac{1}{2}\gamma^2 C^2 + \frac{1}{2}\xi^2(t)\left(B^2 - B\right)\right]\; V(t) + A_t + \lambda\mu\; C = 0\\ &\Rightarrow C_t - \lambda C + \frac{1}{2}\gamma^2 C^2 + \frac{1}{2}\left[B^2 - B\right]\xi^2(t) = 0, \quad A_t + \lambda\mu C = 0, \quad B_t = 0,\\ &\qquad\qquad \textit{with} \qquad A(T) = C(T) = 0, \quad B(T) = i\theta.\end{aligned}$$

So immediately $B(t) = i\theta$, and with the substitution $C = -\frac{2}{\gamma^2}D$

$$\begin{aligned}D_t &= D^2 + \lambda D - \frac{1}{4}\gamma^2\left[i\theta + \theta^2\right]\,\xi^2(t), \quad D(T) = 0,\\ A_t &= \frac{2\lambda\mu}{\gamma^2}D(t), \quad A(T) = 0.\end{aligned}$$

This non-linear ODE for $D(t)$ has an analytic solution only when $\xi(t)$ is constant. So assume $\xi(t) = \xi_j$ is piecewise constant on $(t_j, t_{j+1}]$, and solve for $D(t)$ analytically by backward induction using the solution on each interval as the boundary value for the next iteration. For $t \in (t_j, t_{j+1}]$, set

$$\begin{aligned}D_t &= D^2 + \lambda D - \frac{1}{4}\gamma^2\left[i\theta + \theta^2\right]\,\xi_j^2, \qquad \Delta_j = \lambda^2 + \gamma^2\left[i\theta + \theta^2\right]\,\xi_j^2,\\ d_j^+ &= \frac{1}{2}\left(-\lambda + \sqrt{\Delta_j}\right), \qquad d_j^- = \frac{1}{2}\left(-\lambda - \sqrt{\Delta_j}\right),\end{aligned}$$

and identify two cases depending on whether or not Δ_j is zero:

16.3.1.1 Case $\Delta \neq 0$

Using partial fractions, on $t \in (t_j, t_{j+1}]$

$$\sqrt{\Delta_j}\,(t_{j+1}-t) = \int_t^{t_{j+1}} \frac{dD}{D-d_j^+} - \int_t^{t_{j+1}} \frac{dD}{D-d_j^-}$$
$$= \ln\left(\frac{D-d_j^-}{D-d_j^+}\frac{D_{j+1}-d_j^+}{D_{j+1}-d_j^-}\right)$$

$$\textit{giving} \qquad D(t) = \frac{d_j^+ - d_j^-\left[\frac{D_{j+1}-d_j^+}{D_{j+1}-d_j^-}\exp\left(-\sqrt{\Delta_j}\,(t_{j+1}-t)\right)\right]}{1-\frac{D_{j+1}-d_j^+}{D_{j+1}-d_j^-}\exp\left(-\sqrt{\Delta_j}\,(t_{j+1}-t)\right)},$$

$$\textit{and then} \quad A(t) = A_{j+1} - \frac{2\lambda\mu}{\gamma^2}\frac{1}{\sqrt{\Delta_j}}\int_t^{t_{j+1}} \frac{d_j^+ - d_j^- y}{(1-y)\,y}dy,$$
$$= A_{j+1} - \frac{2\lambda\mu}{\gamma^2}\left\{\frac{d_j^+}{\sqrt{\Delta_j}}\ln\frac{y(t_{j+1})}{y(t)} - \ln\frac{y(t_{j+1})-1}{y(t)-1}\right\},$$

$$\textit{where} \qquad y(t) = \frac{D_{j+1}-d_j^+}{D_{j+1}-d_j^-}\exp\left(-\sqrt{\Delta_j}\,(t_{j+1}-t)\right).$$

16.3.1.2 Case $\Delta = 0$

In this case integrating directly on $t \in (t_j, t_{j+1}]$

$$(t_{j+1}-t) = \int_t^{t_{j+1}} \frac{dD}{\left(D-d_j^+\right)^2} = \frac{1}{D-d_j^+} - \frac{1}{D_{j+1}-d_j^+}, \quad \textit{giving}$$

$$D(t) = \frac{D_{j+1} + d_j^+\left(D_{j+1}-d_j^+\right)(t_{j+1}-t)}{1+\left(D_{j+1}-d_j^+\right)(t_{j+1}-t)}, \qquad \textit{and then}$$

$$A(t) = A_{j+1} + \frac{2\lambda\mu}{\gamma^2}\frac{1}{\left(D_{j+1}-d_j^+\right)}\int_t^{t_{j+1}} \frac{D_{j+1}+d_j^+ y}{1+y}dy,$$
$$= A_{j+1} - \frac{2\lambda\mu}{\gamma^2}\left\{d_j^+(t_{j+1}-t) + \ln\left[1+\left(D_{j+1}-d_j^+\right)(t_{j+1}-t)\right]\right\}.$$

Hence both D and A can be found by backward induction from T to t using $A(T) = D(T) = 0$, yielding (recall $V(0) = 1$)

$$\widehat{p}(\theta;t) = \exp\left(A(t) - \frac{2}{\gamma^2}D(t)\,V(t) + i\theta S(t)\right),$$
$$\textit{and} \quad \widehat{p}(\theta;0) = \exp\left(A(0) - \frac{2}{\gamma^2}D(0) + i\theta S(0)\right).$$

16.3.2 Option price as a Fourier integral

Letting $k = \ln(\kappa + a)$ be the log-shifted strike and $p(s) = p(s, T; X(0), 0)$ be the transition density of $S(T)$ given $X(0)$, then the time $t = 0$ option price $\mathbf{H}(k) = \mathbf{H}(0, \kappa)$ is

$$\mathbf{H}(k) = \mathbf{E}\left\{ [K(T) - \kappa]^{+} \middle| \; X(0) \right\} = \int_k^{\infty} (\exp s - \exp k)\, p(s)\, ds.$$

Our first approach to deriving $\mathbf{H}(k)$ as a Fourier integral is standard and included for completeness, but the second is numerically superior.

16.3.2.1 First Solution

The result follows if for real c, we can find

$$U(k, c) = \int_k^{\infty} p(s) \exp(cs)\, ds \quad \Rightarrow \quad \mathbf{H}(k) = U(k; 1) - \exp k\, U(k; 0).$$

$U(\cdot)$'s Fourier transform $\widehat{U}(\cdot)$, and its inverse back to $U(\cdot)$ are given by

$$\begin{aligned}
\widehat{U}(x, c) &= \int_{-\infty}^{\infty} \left(\int_k^{\infty} p(s) \exp(cs)\, ds \right) \exp(2\pi i x k)\, dk, \\
&= \int_{-\infty}^{\infty} p(s) \exp(cs) \left[\int_{-\infty}^{s} \exp(2\pi i x k)\, dk \right] ds, \\
&= \int_{-\infty}^{\infty} p(s) \exp(cs) \left(\frac{1}{2}\delta(x) + \frac{\exp(2\pi i x s)}{2\pi i x} \right) ds; \\
U(k, c) &= \int_{-\infty}^{\infty} \widehat{U}(x, c) \exp(-2\pi i x k)\, dx, \\
&= \left\{ \begin{array}{c} \frac{1}{2} \int_{-\infty}^{\infty} p(s) \exp(cs)\, ds \\ + \int_{-\infty}^{\infty} \left[\int_{-\infty}^{\infty} p(s) \exp(i [2\pi x - ic] s)\, ds \right] \frac{\exp(-2\pi i x k)}{2\pi i x} dx \end{array} \right\}, \\
&= \frac{1}{2} \int_{-\infty}^{\infty} p(s) \exp(cs)\, ds + \int_{-\infty}^{\infty} \frac{\exp(-2\pi i x k)}{2\pi i x} \widehat{p}(2\pi x - ic)\, dx, \\
&= \frac{1}{2} \int_{-\infty}^{\infty} p(s) \exp(cs)\, ds + \operatorname{Re} \int_0^{\infty} \frac{\exp(-2\pi i x k)}{\pi i x} \widehat{p}(2\pi x - ic)\, dx.
\end{aligned}$$

The last step, halving the integration range and taking real parts, is because

$$\int_{-\infty}^{\infty} \frac{\exp(-2\pi i x k)}{2\pi i x} \widehat{p}(2\pi x - ic)\, dx = \int_0^{\infty} \left\{ \begin{array}{c} \frac{\exp(-2\pi i x k)}{2\pi i x} \widehat{p}(2\pi x - ic) \\ + \frac{\exp(2\pi i x k)}{-2\pi i x} \widehat{p}(-2\pi x - ic) \end{array} \right\} dx,$$

$$\textit{while} \quad \widehat{p}(2\pi x - ic; t) = \mathbf{E}\{ \exp(i [2\pi x - ic] S(T)) \mid X(t) \} \quad \textit{implies}$$

$$\overline{\widehat{p}(2\pi x - ic; t)} = \mathbf{E}\{ \exp(-i [2\pi x + ic] S(T)) \mid X(t) \} = \widehat{p}(-2\pi x - ic).$$

Hence the call option value is given by the Fourier integral

$$\mathbf{H}(k) = \frac{1}{2}[K(T) - \kappa] + \int_0^\infty \mathrm{Re}\left[\frac{1}{\pi i x}\begin{pmatrix} \exp(-2\pi i x k)\,\widehat{p}(2\pi x - i) \\ -\exp(k - 2\pi i x k)\,\widehat{p}(2\pi x) \end{pmatrix}\right] dx.$$

The numerical shortcomings in this expression for $\mathbf{H}(k)$ include:

1. The integrand is singular at $x = 0$.
2. Technically $U(k, c)$ is not integrable in k because $U(-\infty, c) \neq 0$

$$\lim_{k \to -\infty} U(k; 1) = K(0) + a, \qquad \lim_{k \to -\infty} U(k; 0) = \int_{-\infty}^{\infty} p(s)\, ds = 1.$$

3. $U(k, c)$ integrates over a discontinuity, contributing to inaccuracy.

16.3.2.2 Second Solution

Introduce a modified $\mathbf{H}(k)$; for real $b > 0$ define

$$\mathbf{H}_b(k) = \exp(bk)\,\mathbf{H}(k),$$

and note that $\mathbf{H}_b(k)$ now satisfies the growth condition at infinity because

$$\lim_{k \to \infty} \mathbf{H}(k) = 0, \quad \lim_{k \to -\infty} \mathbf{H}(k) = K(0) + a \quad \Rightarrow \quad \lim_{|k| \to \infty} \mathbf{H}_b(k) \to 0.$$

$\mathbf{H}_b(\cdot)$'s Fourier transform $\widehat{\mathbf{H}}_b(\cdot)$, and inverse back to $\mathbf{H}_b(k)$, are given by

$$\widehat{\mathbf{H}}_b(x) = \int_{-\infty}^{\infty} \exp(bk)\left(\int_k^\infty p(s)(\exp s - \exp k)\, ds\right)\exp(2\pi i x k)\, dk,$$

$$= \int_{-\infty}^{\infty}\left\{p(s)\int_{-\infty}^{s}[\exp(s + bk + 2\pi i x k) - \exp(k + bk + 2\pi i x k)]\, dk\right\} ds,$$

$$= \int_{-\infty}^{\infty} p(s)\exp(s + bs + 2\pi i x s)\left(\frac{1}{b + 2\pi i x} - \frac{1}{1 + b + 2\pi i x}\right) ds,$$

$$= \frac{\widehat{p}(2\pi x - i(1+b),\ 0)}{b + b^2 - 4\pi^2 x^2 + 2\pi i x(2b+1)};$$

$$\mathbf{H}(k) = \exp(-bk)\int_{-\infty}^{+\infty} \widehat{\mathbf{H}}_b(x)\exp(-2\pi i k x)\, dx,$$

$$= 2\exp(-bk)\int_0^{+\infty} \mathrm{Re}\,\widehat{\mathbf{H}}_b(x)\exp(-2\pi i k x)\, dx.$$

As long as $b > 0$, there is no singularity in $\widehat{\mathbf{H}}_b(x)$, and because

$$\widehat{p}(2\pi x - i(1+b); 0) = \exp\left[A(0) - \frac{2}{\gamma^2}D(0) + (1+b)S(0)\right]\exp[i2\pi x S(0)]$$

it should stay bounded for finite b (a satisfactory default value for b is $b = \frac{1}{2}$).

16.4 Simulation

The Glasserman approach of Section-9.1 translates well to stochastic volatility BGM. Relationships between his $Z(t,T)$ and $V(t,T)$ variables (please distinguish between Glasserman's $V(t,T)$ and the stochastic variance $V(t) = V_t$), forwards $K(\cdot)$ and bonds $B(\cdot)$ remain the same, but, as above, there is an extra $\sqrt{V(t)}$ term in volatilities that will require simulation.

Under the *terminal measure* $\mathbb{P}_n$ clearly $Z(t,T_n) = 1$ and SDEs for the $Z(t,T)$ for $j = @(t), .., n-1$ are

$$\begin{aligned}\frac{dZ(t,T_j)}{Z(t,T_j)} &= \sqrt{V(t)}b^*(t,T_j,T_n)\,dW_n(t)\\ &= \sqrt{V(t)}\left[\sum_{\ell=j}^{n-1}\phi\left\{1-\lambda_\ell\frac{Z(t,T_{\ell+1})}{Z(t,T_\ell)}\right\}\xi^*(t,T_\ell)\right]dW_n(t),\end{aligned}$$

while $V(t,T_n) = 1$ and SDEs for the $V(t,T)$ for $j = @(t), .., n-1$ are

$$\begin{aligned}\frac{dV(t,T_j)}{V(t,T_j)} &= \sqrt{V(t)}\left[\xi^*(t,T_j)+\sum_{\ell=j+1}^{n-1}\phi\left\{\frac{V(t,T_\ell)}{Z(t,T_\ell)}\right\}\xi^*(t,T_\ell)\right]dW_n(t),\\ Z(t,T_\ell) &= V(t,T_\ell)+\Pi_\ell^\ell V(t,T_{\ell+1})\ldots+\ \Pi_\ell^{n-1}V(t,T_n),\end{aligned}$$

in which $\sum_{\ell=n}^{n-1}[\cdot] = 0$ when $j = n-1$.

Under the *spot measure* $\mathbb{P}_0$ SDEs for the $Z(\cdot)$ are $dZ(t,T_@) = 0$ and for $j = @(t)+1, .., n$

$$\begin{aligned}\frac{dZ(t,T_j)}{Z(t,T_j)} &= -\sqrt{V(t)}b^*(t,T_@,T_j)\,dW_0(t)\\ &= -\sqrt{V(t)}\left[\sum_{\ell=@(t)}^{j-1}\phi\left\{\frac{V(t,T_\ell)}{Z(t,T_\ell)}\right\}\xi^*(t,T_\ell)\right]dW_0(t),\end{aligned}$$

while corresponding SDEs for the $V(\cdot)$ for $j = @(t), .., n$ are

$$\begin{aligned}\frac{dV(t,T_j)}{V(t,T_j)} &= \sqrt{V(t)}\left[\xi^*(t,T_j)-\sum_{\ell=@(t)}^{j}\phi\left\{\frac{V(t,T_\ell)}{Z(t,T_\ell)}\right\}\xi^*(t,T_\ell)\right]dW_0(t),\\ Z(t,T_\ell) &= V(t,T_\ell)+\Pi_\ell^\ell V(t,T_{\ell+1})\ldots+\ \Pi_\ell^{n-1}V(t,T_n).\end{aligned}$$

16.4.1 Simulating $V(t)$

The behavior of *square-root diffusions* like our variance process $V(t)$

$$dV(t) = \lambda[\mu - V(t)]\,dt + \gamma\sqrt{V(t)}dU(t), \quad \lambda,\mu,\gamma > 0 \quad V(0) = 1,$$

is well known, see Glasserman's book [46] Section-3.4 for background and references. For example, the mean and long-term mean of $V(t)$ are given by

$$\frac{d\mathbf{E}V(t)}{dt} = \lambda\mu - \lambda\mathbf{E}V(t) \quad \Rightarrow \quad \mathbf{E}V(t) = \mu + (V(0) - \mu)\exp(-\lambda t) \to \mu$$

while, after Ito plus manipulation, the variance and long-term variance are

$$d[V(t)]^2 = 2V(t)\left[\lambda(\mu - V(t))\,dt + \gamma\sqrt{V(t)}dW_t\right] + \gamma^2 V(t)\,dt$$

$$\mathrm{Var}\,V(t) = \mathbf{E}[V(t)]^2 - [\mathbf{E}V(t)]^2 = \frac{\gamma^2}{2\lambda}\left\{\begin{array}{c} 2V(0)\left(\mathbf{e}^{-\lambda t} - \mathbf{e}^{-2\lambda t}\right) \\ +\mu\left(1 - \mathbf{e}^{-\lambda t}\right)^2 \end{array}\right\} \to \frac{\gamma^2\mu}{2\lambda}.$$

More specifically $V(t)$ given $V(u)$ behaves like the random variable

$$V(t) = \frac{\gamma^2\left(1 - \mathbf{e}^{-\lambda(t-u)}\right)}{4\lambda}\,\chi^2_\nu\left(\frac{4\lambda\mathbf{e}^{-\lambda(t-u)}}{\gamma^2\left(1 - \mathbf{e}^{-\lambda(t-u)}\right)}V(u)\right) \qquad \nu = \frac{4\mu\lambda}{\gamma^2},$$

where $\chi^2_\nu(\lambda)$ is a *non-central chi-square* random variable with *non-centrality parameter* λ and ν *degrees of freedom* (which need not be an integer). Glasserman gives detailed methods of simulating $V(t)$ whatever the (positive) values of λ, μ and γ, that is, whether the degree of freedom ν is integral or not.

But if ν is an integer, the square root process $V(t)$ can be expressed as the sum of squares of ν independent OU processes $X_j(t)$ as follows. Starting with

$$\begin{aligned}
dX_j(t) &= -\frac{1}{2}\lambda\,X_j(t)\,dt + \frac{1}{2}\gamma dW_i(t), \qquad j = 1,..,\nu \\
\textit{let} \quad & V(t) = X_1^2(t) + X_2^2(t) + .. + X_\nu^2(t), \qquad \Rightarrow \\
dV(t) &= \lambda\left[\frac{\gamma^2\nu}{4\lambda} - V(t)\right]dt + \gamma\sum_{j=1}^{\nu} X_j(t)\,dW_i(t), \quad \Rightarrow \\
dV(t) &= \lambda[\mu - V(t)]\,dt + \gamma\sqrt{V(t)}d\widetilde{W}(t), \\
\textit{where} \quad d\widetilde{W}(t) &= \sum_{j=1}^{\nu}\frac{X_j(t)}{\sqrt{V(t)}}dW_i(t) \quad \textit{and} \quad \mu = \frac{\gamma^2\nu}{4\lambda}.
\end{aligned}$$

So restrictions on ν can be exchanged for the relative ease of simulating several OU processes (rather than Gamma and Poisson distributions, see [46] Section-3.4). In practice $\nu = 3$, or somewhat larger, seems the only reasonable choice, as $\nu = 1$ or $\nu = 2$ make $V(t) = 0$ attainable (see Section-A.6).

REMARK 16.2 Because rapid mean reversion may be required to get a suitable convexity profile, the time steps used to simulate $V(t)$ may need to be quite small (for methods of taking large accurate steps see Broadie and Kaya [34]). Note that if the stochastic volatility parameters do not change frequently, simulated trajectories of $V(t)$ can be prepared at leisure, tabulated and reused, because they are independent of the yieldcurve drivers. □

16.5 Interpolation, Greeks and calibration

We very briefly mention some of the practical problems in making stochastic BGM operational.

16.5.1 Interpolation

As in Section-8.1, for $0 \leq t \leq T_0 < T < T_1$ suppose $K(t, T_0)$ and $K(t, T_1)$ are known and $K(t, T)$ is required. Interpolate $\xi(t, T)$ on maturity T, defining

$$\begin{aligned} \xi(t,T) &= \frac{1}{\delta}\theta(T)\left\{(T_1 - T)\,\xi(t,T_0) + (T - T_0)\,\xi(t,T_1)\right\}, \\ &= \alpha(T)\,\xi(t,T_0) + \beta(T)\,\xi(t,T_1), \\ \textit{with} \quad & \theta(T_0) = 1 = \theta(T_1), \end{aligned}$$

which preserves correlation between forwards at nodepoints and allows $\theta(\cdot)$ to be chosen to satisfy some auxiliary condition. Recall

$$\ln \frac{K(t,T_0)}{K(0,T_0)} = \int_0^t \sqrt{V(s)}\xi^*(s,T_0)\,dW_{T_1}(s) - \frac{1}{2}\int_0^t V(s)\,|\xi(s,T_0)|^2\,ds,$$

so if the forward measures $\mathbb{P}_T$ and $\mathbb{P}_{T_1}$ are identified with $\mathbb{P}_{T_0}$, clearly

$$\begin{aligned} \ln \frac{K(t,T)}{K(0,T)} &\cong \alpha(T) \ln \frac{K(t,T_0)}{K(0,T_0)} + \beta(T) \ln \frac{K(t,T_1)}{K(0,T_1)} \\ &+ \frac{1}{2}\int_0^t V(s)\,\left[\alpha(T)\,|\xi(s,T_0)|^2 + \beta(T)\,|\xi(s,T_1)|^2 - |\xi(s,T)|^2\right] ds. \end{aligned}$$

The convexity term in this interpolation for $K(t,T)$ in terms of $K(t,T_0)$ and $K(t,T_1)$ is now, however, stochastic and must either be jointly simulated with the Glasserman $V(t,T)$ variable or made deterministic by approximating the variance process $V(s)$ with its expected value $\mathbf{E}V(s) = \mu + (V(0) - \mu)\exp(-\lambda s)$ or long-term mean μ.

16.5.2 Greeks

Assume that the stochastic volatility parameters λ, μ, γ have been fixed to return a desired convexity profile along the caplet or swaption implied volatility surface, and they remain stable for a period of several months so they do not have to be hedged (though reserves against possible longer term changes may be needed).

With the stochastic parameters λ, μ, γ constant, it is easy to see that the pathwise delta and vega methods already developed for shifted BGM in Chapters 11 and 13 work well with stochastic volatility BGM.

For *pathwise deltas*, the relevant spot measure equation (11.2) to be partially differentiated becomes

$$\frac{H(t,T_j)}{H(0,T_j)} = \mathcal{E}\left\{\int_0^t \mu_j^{(0)}(t)\,dt + \int_0^t \sqrt{V_t}\xi^*(t,T_i)\,dW_0(t)\right\},$$

$$\textit{where} \qquad \mu_j^{(0)}(t) = V_t\xi^*(t,T_j) \sum_{j1=@}^{j} h_{j1}(0)\,\xi(t,T_{j1}),$$

while the equivalent equation (11.3) under the terminal measure becomes

$$\frac{H(t,T_j)}{H(0,T_j)} = \mathcal{E}\left\{\int_0^t \mu_j^{(n)}(t)\,dt + \int_0^t \sqrt{V_t}\xi^*(t,T_i)\,dW_n(t)\right\},$$

$$\textit{where} \qquad \mu_j^{(n)}(t) = -V_t\xi^*(t,T_j) \sum_{j1=j+1}^{n-1} h_{j1}(0)\,\xi(t,T_{j1}).$$

In either case, partial differentiation of $H(t,T_j)$ with respect to $H(0,T_\ell)$, or equivalently $K(0,T_\ell)$ proceeds similarly to (11.4), while subsequent formulae for partial derivatives of swaps and options in terms of the forwards in Chapter-11 remain the same.

For *vegas*, in (16.5), separately perturb by small $\Delta\theta$ the shift $\alpha(T_j,T_N)$

$$\alpha(T_j,T_N) \quad \rightarrow \quad (1+\Delta\theta)\,\alpha(T_j,T_N),$$

and by small $\Delta\varepsilon$ the instantaneous volatility $\sum_{j=0}^{N-1} A_j\xi(t,T_j)$

$$\sum_{j=0}^{N-1} A_j\xi(t,T_j) \quad \rightarrow \quad (1+\Delta\varepsilon)\sum_{j=0}^{N-1} A_j\xi(t,T_j)$$

of just the jth swaption $\mathrm{pSwpn}(t,\kappa,T_j,T_N)$. Corresponding changes in the swaption value are given by

$$\Delta_\theta\,\mathrm{pSwpn}(0,\kappa,T_j,T_N) = \mathrm{pSwpn}((1+\Delta\theta)\,\alpha(T_j,T_N)) - \mathrm{pSwpn}(\alpha(T_j,T_N)),$$

$$\Delta_\varepsilon\,\mathrm{pSwpn}(0,\kappa,T_j,T_N) = \left\{\begin{array}{c}\mathrm{pSwpn}\left((1+\Delta\varepsilon)\sum_{j=0}^{N-1} A_j\xi(t,T_j)\right)\\ -\,\mathrm{pSwpn}\left(\sum_{j=0}^{N-1} A_j\xi(t,T_j)\right)\end{array}\right\},$$

in which the swaption values $\mathrm{pSwpn}(\cdot)$ must now of course be computed by the Heston option methods of Section-16.3. Afterward proceed as in Chapter-13.

16.5.3 Caplet calibration

On the understanding that λ,μ,γ are fixed, we now outline a procedure to fit at-the-money caplet volatility and skew on a daily basis. Start by fitting a shifted BGM model (that is, one without stochastic volatility) to the at-the-money volatilities $\beta(T)$ and the at-the-money skews $\partial\beta(T)$ of the *target* caplet implied volatility surface to obtain a shift $a(T)$ and forward volatility $\xi(t,T)$ calibration. If we perturb these functions in the following way that affects only the T_j-maturing caplet

$$a(T_j) \to (1+\varepsilon)\ a(T_j) \qquad \xi(t,T_j) \to (1+\theta)\ \xi(t,T_j),$$

then clearly increasing ε increases the caplet's at-the-money skew, while increasing θ increases its at-the-money volatility, and vice-versa.

Step-1: Using $a(T)$, $\xi(t,T)$ and the fixed λ, μ, γ already determined, find the resultant *secondary* caplet implied volatility surface with at-the-money volatility $\beta_1(T)$ and the at-the-money skew $\partial\beta_1(T)$.

Step-2: If $\partial\beta_1(T_j) > \partial\beta(T_j)$ reduce ε, but increase it if $\partial\beta_1(T_j) < \partial\beta(T_j)$, and if $\beta_1(T_j) > \beta(T_j)$ reduce θ, but increase it if $\beta_1(T_j) < \beta(T_j)$.

Step-3: Iterate by returning to Step-1 with new perturbed functions $a(T)$ and $\xi(t,T)$ but the same fixed λ, μ, γ.

16.5.4 Swaption calibration

The process is similar to that for caplets. Start by fitting a shifted BGM model to the at-the-money volatilities $\beta(T,T_N)$ and the at-the-money skews $\partial\beta(T,T_N)$ of the *target* swaption implied volatility surface to obtain a shift $a(T)$ and forward volatility $\xi(t,T)$ calibration. If we perturb these functions using the technique developed for computing swaption vegas in Chapter-13 that affects only the T_j-maturing swaption shift and implied volatility but leaves others unchanged, that is

$$\alpha(T_j,T_N) \to (1+\varepsilon)\ \alpha(T_j,T_N) \qquad \sigma(t,T_j,T_N) \to (1+\theta)\ \sigma(t,T_j,T_N),$$

then clearly increasing ε increases the swaption's at-the-money skew, while increasing θ increases its at-the-money volatility, and vice-versa.

Step-1: Using $a(T)$, $\xi(t,T)$ and the fixed λ, μ, γ already determined, find the resultant *secondary* swaption implied volatility surface with at-the-money volatilities $\beta_1(T,T_N)$ and the at-the-money skews $\partial\beta_1(T,T_N)$.

Step-2: If $\partial\beta_1(T_j,T_N) > \partial\beta(T_j,T_N)$ reduce ε, but increase it if $\partial\beta_1(T,T_N) < \partial\beta(T,T_N)$, and if $\beta_1(T,T_N) > \beta(T,T_N)$ reduce θ, but increase it if $\beta_1(T,T_N) < \beta(T,T_N)$.

Step-3: With the perturbed swaption shifts $\alpha(T,T_N)$ and volatilities $\sigma(t,T,T_N)$ compute the new perturbed $a(T)$ and $\xi(t,T)$ using the vega 'inversion technique'.

Step-4: Iterate by returning to Step-1 with new perturbed functions $a(T)$ and $\xi(t,T)$ but the same fixed λ, μ, γ.

Chapter 17

Options in Brazil

This *extra chapter,* which addresses some of the theoretical aspects of Brazilian options, can be justified by some BGM applications, but has been added mostly because the content appeals to the author! At the outset, it should be emphasized that it may contain mistakes and biased interpretations due to the author's relative unfamiliarity with the area (details of some instruments are a struggle, in particular, the system of contract dates).

For more information, readers might like to refer to the English language website [26] for the interest rate section of the *Bolsa de Mercadorias & Futuros* (BMF), which is the main Brazilian Mercantile & Futures Exchange.

17.1 Overnight DI

Without ever having lived in Brazil, the author nevertheless imagines that frequent financial crises combined with a relatively benign (for South America) government attitude to financial markets, have naturally lead to the invention of safe and flexible ways of increasing, or at least maintaining, wealth over short time horizons.

For that reason he is not surprised that the foundation for most Brazilian interest rate derivatives is the *CDI rate* or *overnight DI (Deposito Interbancario) rate* $D(t)$, which is an annualized rate paying

$$[1 + D(t)]^{\frac{1}{252}}$$

over a one day period at time t. It is calculated and published daily, and represents the average rate of all inter-bank overnight transactions in Brazil: banks usually express their cost of funding as a percentage of the published CDI terms.

The *IDI index* $\mathrm{IDI}(t)$ accumulates those daily payments from some start date

$$\mathrm{IDI}(t) = \prod_{i=0}^{n-1} [1 + D(t_i)]^{\frac{1}{252}}, \qquad n = \lfloor 252 \times t \rfloor$$

resetting to 100,000 from time-to-time. The IDI index therefore behaves like a bank account, and so it can be reasonably modelled in HJM style by setting

$$[1+D(t)]^{\frac{1}{252}} = 1 + r(t)\,\Delta(t) \cong \exp r(t)\,\Delta(t), \tag{17.1}$$
$$\mathrm{IDI}(t) = \beta(t) = \exp \int_0^t r(s)\,ds,$$

where $\Delta(t)$ is one day and $r(t)$ is the spot interest rate, or, in BGM fashion, by rolling it up into consecutive zero coupon bonds that mature daily.

17.2 Pre-DI swaps and swaptions

These are standard over-the-counter deals tailored to the needs of the counterparties, with the convention (the author's understanding) that the length of swaps is always a whole number of months (that is, deals begin and end on the same day of the month).

By definition the *Pre-DI payer swap* struck at K and accumulated over the interval $[T, T_1]$, where $T_1 = \delta + T$, pays

$$\frac{\mathrm{IDI}(T_1)}{\mathrm{IDI}(T)} - K = \frac{\beta(T_1)}{\beta(T)} - K$$

at time T_1. The time t *swaprate* $\omega(t) = \omega(t, T, T_1)$ is that value of K which makes the swap's time t value zero

$$\mathbf{E}_0 \left\{ \frac{1}{\beta(T_1)} \left[\frac{\beta(T_1)}{\beta(T)} - \omega(t) \right] \middle| \mathcal{F}_t \right\} = 0,$$

namely

$$\omega(t) = \frac{B(t,T)}{B(t,T_1)} = \frac{1}{F_T(t,T_1)},$$

where $\mathbf{E}_0$ is expectation under the spot measure $\mathbb{P}_0$, $B(t,T)$ is the time t value of a zero coupon bond maturing at time T, and $F_T(t,T_1)$ is a T-forward contract on the zero maturing at T_1.

To make numbers comparable with the CDI rate, the convention is that the actual market *quoted swaprate* $f(t,T,T_1)$ and actual market *quoted strike* κ are obtained from $\omega(t,T,T_1)$ and K by the unique one-to-one formulae

$$\omega(t) = \omega(t,T,T_1) = [1 + f(t,T,T_1)]^{\frac{T_1 - T}{252}}, \qquad K = [1+\kappa]^{\frac{T_1 - T}{252}}. \tag{17.2}$$

Also, to develop models, we will need the notion of *forward accrual* $A(t)$ over the interval $[T, T_1]$

$$A(t) = A(t,T,T_1) = [1 + f(t,T,T_1)]^{\frac{T_1 - T}{252}} - 1 \tag{17.3}$$
$$= \omega(t) - 1 = \frac{1}{F_T(t,T_1)} - 1,$$

which is virtually (give or take a scaling factor) the simple forward over $[T, T_1]$.

Substituting back, the time t value of a *payer swap* (paying fixed and receiving floating) is therefore

$$\begin{aligned}
&\text{pSwap}(t) = \text{pSwap}(t, T, T_1) \qquad (17.4)\\
&= \mathbf{E}_0 \left\{ \frac{1}{\beta(T_1)} \left[\frac{\beta(T_1)}{\beta(T)} - K \right] \middle| \mathcal{F}_t \right\} = \mathbf{E}_0 \left\{ \frac{1}{\beta(T_1)} \left[\omega(t) - K \right] \middle| \mathcal{F}_t \right\},\\
&= B(t, T_1) \left[\omega(t) - K \right] = B(t, T) - K\ B(t, T_1),\\
&= B(t, T_1) \left[A(t) - (K - 1) \right],
\end{aligned}$$

which has exactly the same form as any standard swap, except that there is just one exchange.

The time t value of the corresponding *payer swaption* can be expressed in several ways

$$\begin{aligned}
&\text{pSwpn}(t) = \text{pSwpn}(t, T, T_1) \qquad (17.5)\\
&= \beta(t)\, \mathbf{E}_0 \left\{ \frac{B(T, T_1)}{\beta(T)} \left[\omega(T) - K \right]^+ \middle| \mathcal{F}_t \right\}\\
&= B(t, T)\, \mathbf{E}_T \left\{ \left[1 - K F_T(T, T_1) \right]^+ \middle| \mathcal{F}_t \right\},\\
&= B(t, T_1)\, \mathbf{E}_{T_1} \left\{ \left[A(T) - (K - 1) \right]^+ \middle| \mathcal{F}_t \right\},
\end{aligned}$$

where $\mathbf{E}_T$ (respectively $\mathbf{E}_{T_1}$) is expectation under the T-forward measure $\mathbb{P}_T$ (respectively $\mathbb{P}_{T_1}$). Similar expressions hold for *receiver swaptions*

$$\begin{aligned}
\text{rSwap}(t) &= \text{rSwap}(t, T, T_1) = K\ B(t, T_1) - B(t, T), \qquad (17.6)\\
&= B(t, T_1) \left[(K - 1) - A(t) \right],\\
\text{rSwpn}(t) &= \text{rSwpn}(t, T, T_1) = B(t, T)\, \mathbf{E}_T \left\{ \left[K F_T(T, T_1) - 1 \right]^+ \middle| \mathcal{F}_t \right\},\\
&= B(t, T_1)\, \mathbf{E}_{T_1} \left\{ \left[(K - 1) - A(T) \right]^+ \middle| \mathcal{F}_t \right\}.
\end{aligned}$$

Note that these results are *model independent*, and that the options can be regarded as either caps/floors or swaptions because they are on swaps with just one exchange.

17.2.1 In the HJM framework

An SDE for the forward $F_T(t,T_1)$ under $\mathbb{P}_T$ is

$$dF_T(t,T_1) = -F_T(t,T_1)\int_T^{T_1}\sigma^*(t,u)\,du\;dW_T(t) \quad\Rightarrow$$

$$F_T(T,T_1) = F_T(t,T_1)\,\mathcal{E}\left\{\int_t^T\int_T^{T_1}\sigma^*(s,u)\,du\;dW_T(s)\right\},$$

giving, from (17.5), (17.6) and the Black $\mathbf{B}(\cdot)$ formula (A.2.3), the following HJM style swaption formulae

$$\begin{aligned}\text{pSwpn}(t) &= B(t,T)\,\mathbf{B}(1,\ K\ F_T(t,T_1),\ \zeta), \qquad (17.7)\\ \text{rSwpn}(t) &= B(t,T)\,\mathbf{B}(K\ F_T(t,T_1),\ 1,\ \zeta),\\ \zeta^2 &= \int_t^T\left|\int_T^{T_1}\sigma(s,u)\,du\right|^2 ds.\end{aligned}$$

Note that in the flat Ho & Lee case that produces a Black option formula with implied volatility $(T_1 - T)\ \sigma$.

17.2.2 In the BGM framework

The SDE for the reciprocal of the forward under $\mathbb{P}_{T_1}$ yields the following SDE for the accrual $A(t)$

$$d\left(\frac{1}{F_T(t,T_1)}\right) = \left(\frac{1}{F_T(t,T_1)}\right)\int_T^{T_1}\sigma^*(t,u)\,du\,dW_{T_1}(t) \quad\Rightarrow$$

$$\frac{dA(t,T,T_1)}{1+A(t,T,T_1)} = \gamma^*(t,T,T_1)\ dW_{T_1}(t),$$

if, in BGM style, the stochastic HJM volatility is chosen to satisfy

$$(1+A(t,T,T_1))\int_T^{T_1}\sigma(t,u)\,du = A(t,T,T_1)\,\gamma(t,T,T_1),$$

where $\gamma(t,T,T_1)$ is deterministic. From (17.5), (17.6) and the Black $\mathbf{B}(\cdot)$ formula (A.2.3), that gives the following BGM style formulae for the swaptions

$$\begin{aligned}\text{pSwpn}(t) &= B(t,T_1)\,\mathbf{B}(A(t,T,T_1),\ K-1,\ \zeta), \qquad (17.8)\\ \text{rSwpn}(t) &= B(t,T_1)\,\mathbf{B}(K-1,\ A(t,T,T_1),\ \zeta),\\ \zeta^2 &= \int_t^T|\gamma(s,T,T_1)|^2\,ds.\end{aligned}$$

17.3 DI index options

These are an exchange traded options, maturing at the beginning of the months Jan, April, July, Oct and also the month following the current month. But as we will see, the volatilities of these options steadily decrease (to zero at maturity), which seems to make them difficult to use in practice; consequently they are neither heavily traded nor used to hedge swaptions.

The payoff at time T for a DI Index option is the accumulated index starting from some reference time T^*, which is generally take to be zero. Hence DI call and put options struck at K will respectively have time t value

$$\begin{aligned}
\mathrm{DIcall}(t,T) &= \mathbf{E}_0\left\{\frac{\beta(t)}{\beta(T)}[\beta(T)-K]^+\middle|\mathcal{F}_t\right\}, \\
&= \mathbf{E}_0\left\{\left[\beta(t)-K\exp\left(-\int_t^T r(s)\,ds\right)\right]^+\middle|\mathcal{F}_t\right\}. \\
\mathrm{DIput}(t,T) &= \mathbf{E}_0\left\{\left[K\exp\left(-\int_t^T r(s)\,ds\right)-\beta(t)\right]^+\middle|\mathcal{F}_t\right\}.
\end{aligned}$$

Note that the reference time effectively scales the contract, because a later reference time T^* would result in an option payoff of $\frac{\beta(T)}{\beta(T^*)}$ instead of $\beta(T)$.

Getting an option formula is easy in the HJM case, but messy in the BGM framework and not attempted here.

17.3.1 In the HJM framework

An SDE for the zero coupon bond $B(t,T)$ is

$$dB(t,T) = B(t,T)\left[r(t)\,dt - \int_t^T \sigma^*(t,u)\,du\,dW_0(t)\right], \tag{17.9}$$

which has time T solution

$$B(T,T) = 1 = \left\{\begin{array}{c} B(t,T)\exp\left(\int_t^T r(s)\,ds\right) \\ \times\mathcal{E}\left(-\int_t^T\int_s^T \sigma^*(s,u)\,du\,dW_0(s)\right)\end{array}\right\}, \tag{17.10}$$

$$\Rightarrow \quad \exp\left(-\int_t^T r(s)\,ds\right) = B(t,T)\ \mathcal{E}\left(-\int_t^T\int_s^T \sigma^*(s,u)\,du\,dW_0(s)\right).$$

Hence, from (A.2.3), the time t values of the options are

$$\begin{aligned}
\mathrm{DIcall}\,(t,T) &= \mathbf{E}_0\left\{\left[\begin{array}{c}\beta\,(t)\\ -K\;B\,(t,T)\;\;\mathcal{E}\left(-\int_t^T\int_s^T \sigma^*\,(s,u)\,du\;dW_0\,(s)\right)\end{array}\right]^+\middle|\,\mathcal{F}_t\right\},\\
&= \mathbf{B}\,(\beta\,(t)\,,\;K\;B\,(t,T)\,,\;\zeta)\\
\mathrm{DIput}\,(t,T) &= \mathbf{B}\,(K\;B\,(t,T)\,,\;\beta\,(t)\,,\;\zeta)\\
\zeta^2 &= \int_t^T \left|\int_s^T \sigma\,(s,u)\,du\right|^2 ds.
\end{aligned}$$

which are Black-Scholes expressions with volatilities contracting to zero. In the flat Ho & Lee case they yield the corresponding Black-Scholes implied volatility

$$\frac{1}{\sqrt{3}}\,(T-t)\,\sigma.$$

17.4 DI futures contracts

Futures contracts and options on them, are exchange traded instruments that mature at the beginning of the months Jan, April, July, Oct and also the month following the current month; hence underlying contracts may have 1,3,6 or 12 months to run. The options are often used to hedge OTC swaptions, because their dynamics are similar (give or take differences in maturity) and their volatilities well behaved.

The time T maturing futures contract can be entered or exited at any time without cost, its numerical *value* equals that of a zero coupon bond $B\,(t,T)$ maturing at the same time, and the *daily margin payments* $\Delta M_T\,(t)$ associated with it are, from (17.1) with Δt one day,

$$\begin{aligned}
\Delta M_T\,(t) &= B\,(t+\Delta t,T) - B\,(t,T)\,[1+D\,(t)]^{\frac{1}{252}}\,,\\
&= B\,(t+\Delta t,T) - B\,(t,T)\,[1+r\,(t)\,\Delta t]\,.
\end{aligned}$$

In the continuous case the dynamics (17.9) of a zero coupon contract imply

$$\begin{aligned}
dM_T\,(t) &= dB\,(t,T) - B\,(t,T)\,r\,(t)\,dt \qquad (17.11)\\
&= -B\,(t,T)\int_t^T \sigma^*\,(t,u)\,du\;dW_0\,(t)\,,
\end{aligned}$$

that is, the daily margin payment equals the daily change in the stochastic part of the zero coupon bond. Hence the time t value of a DI futures contract (exiting at an arbitrary time $T_1 \le T$) being equal to the present value of the margin payments

$$\mathbf{E}_0 \left\{ \int_t^{T_1} \frac{dM_T(s)}{\beta(s)} \middle| \mathcal{F}_t \right\}$$

$$= -\mathbf{E}_0 \left\{ \int_t^{T_1} \frac{B(s,T)}{\beta(s)} \int_s^T \sigma^*(s,u)\,du\,dW_0(s) \middle| \mathcal{F}_t \right\} = 0,$$

must be zero. This must be true, of course, because of the zero cost of entering and leaving the contract.

REMARK 17.1 Although the DI contract delivers nothing (in contrast to a standard futures contract which delivers something even if cash settled), it is a *futures contract* in the sense that it costs nothing to enter or leave and settles at the margin. Also the margin payments can be duplicated by borrowing to purchase a zero coupon and financing it at the overnight rate. □

An *alternative approach* is to take the given properties of the contract and posit an index $B(t,T)$ which:

1. Is positive and converges to unity at its *maturity* T, that is $B(T,T) = 1$.
2. Can be entered and exited freely with margin payments

   $$dM_T(t) = dB(t,T) - B(t,T)\,r(t)\,dt$$

   that have zero present value.

For the margin payments to have zero present value whatever entry and exit times $M_T(t)$ must be a $\mathbb{P}_0$ martingale (see Section-A.7.1), and so $B(t,T)$ satisfies an SDE like

$$\frac{dB(t,T)}{B(t,T)} = r(t)\,dt + \xi^*(t,T)\,dW_0(t)$$

for some (possibly stochastic) volatility function $\xi(t,T)$. The solution at maturity T

$$B(T,T) = 1 = B(t,T)\exp\left(\int_t^T r(s)\,ds\right)\,\mathcal{E}\left(-\int_t^T \xi^*(s,T)\,dW_0(s)\right),$$

$$\Rightarrow \quad B(t,T)\ \ \mathcal{E}\left(-\int_t^T \xi^*(s,T)\,dW_0(s)\right) = \exp\left(-\int_t^T r(s)\,ds\right),$$

$$\Rightarrow \quad B(t,T) = \mathbf{E}_0\left\{\exp\left(-\int_t^T r(s)\,ds\right) \middle| \mathcal{F}_t\right\},$$

which, according to (1.7), is a zero coupon bond.

17.4.1 Hedging with futures contracts

The futures contract can be used to directly hedge pre-DI swaps, because from (17.4), (17.9) and (17.11)

$$d\,\mathrm{pSwap}\,(t) = \mathrm{pSwap}\,(t)\, r\,(t)\, dt + dM_T\,(t) - K\; dM_{T_1}\,(t)\,,$$

and this result is *independent of the model* (no assumptions made about the HJM volatility which may be stochastic).

In the HJM framework, using (A.3.8) we can hedge swaptions via

$$\begin{aligned} d\,\mathrm{pSwpn}\,(t) = {} & \mathrm{pSwpn}\,(t)\;\; r\,(t)\, dt + \mathbf{B}\,(1,\;\; K\;\, F_T\,(t,T_1)\,,\;\; \zeta)\;\; dM_T\,(t) \\ & + \Delta\mathbf{B}\,(1,\;\; K\;\, F_T\,(t,T_1)\,,\;\; \zeta)\,[dM_{T_1}\,(t) - F_T\,(t,T_1)\;\; dM_T\,(t)]\,, \end{aligned}$$

and DI Index options via

$$d\,\mathrm{DIcall}\,(t,T) = \mathrm{DIcall}\,(t,T)\;\; r\,(t)\, dt + \Delta\mathbf{B}\left(\beta\,(t)\,,\;\; K\;\, B\,(t,T)\,,\;\; \zeta^2\right)\; K\, dM_T\,(t)\,.$$

Similarly *in the BGM framework* we can hedge swaptions via

$$\begin{aligned} d\,\mathrm{pSwpn}\,(t) = {} & \mathrm{pSwpn}\,(t)\;\; r\,(t)\, dt + \mathbf{B}\,(A\,(t,T,T_1)\,,\;\; K-1,\;\; \zeta)\;\; dM_{T_1}\,(t) \\ & + \Delta\mathbf{B}\,(A\,(t,T,T_1)\,,\;\; K-1,\;\; \zeta) \\ & \times [dM_T\,(t) - (1 + A\,(t,T,T_1))\, dM_{T_1}\,(t)]\,. \end{aligned} \tag{17.12}$$

17.5 DI futures options

This is an option on a zero coupon bond maturing at time T_1 with payoffs at time T like

$$\begin{aligned} \mathrm{payer}\,(T) &= \left[B\,(T,T_1) - \frac{1}{[1+\kappa]^{\frac{T_1-T}{252}}}\right]^+ = \left[B\,(T,T_1) - \frac{1}{K}\right]^+, \\ \mathrm{receiver}\,(T) &= \left[\frac{1}{[1+\kappa]^{\frac{T_1-T}{252}}} - B\,(T,T_1)\right]^+ = \left[\frac{1}{K} - B\,(T,T_1)\right]^+, \end{aligned}$$

where $K = [1+\kappa]^{\frac{T_1-T}{252}}$. Hence from (17.3)

$$\begin{aligned} \mathrm{payer}\,(t) &= B\,(t,T_1)\,\mathbf{E}_{T_1}\left\{\frac{1}{B\,(T,T_1)}\left[B\,(T,T_1) - \frac{1}{K}\right]^+\middle|\,\mathcal{F}_t\right\}, \\ &= \frac{1}{K} B\,(t,T_1)\,\mathbf{E}_{T_1}\left\{[(K-1) - A\,(T)]^+\middle|\,\mathcal{F}_t\right\}, \\ &= \frac{1}{K} B\,(t,T_1)\,\mathbf{B}\,(K-1,\;\; A\,(t,T,T_1)\,,\;\; \zeta)\,. \\ \mathrm{receiver}\,(t) &= \frac{1}{K} B\,(t,T_1)\,\mathbf{B}\,(A\,(t,T,T_1)\,,\;\; K-1,\;\; \zeta)\,. \end{aligned}$$

These formulae are almost the same as those for swaptions (17.8), and so hedges as similar to those given by (17.12).

Appendix A

Notation and Formulae

A.1 Swap notation

This is a summary of the notation introduced in Section-2.2

j	floating index $j = 0, 1, 2, ...$
T_j	floating period time nodes
$(T_{j-1}, T_j]$	j^{th} floating side interval
Δ_j	$= T_j - T_{j-1}$ width of jth interval
$K(t, T_j)$	cash forward rate
$L(t, T_j)$	$= K(t, T_j) + \mu_j$ Libor
$H(t, T_j)$	$= K(t, T_j) + b(T_j)$ shifted cash rate
δ_j	$= \delta(T_j) = \Delta_{j+1}$ coverage of $K(t, T_j)$
μ_j	$= \mu(T_j)$ margin for $L(t, T_j)$
N	number floating payments in a swap
$j0, .., jN$	floating side index inside a swap
i	fixed index $i = 0, 1, 2, ...$
$\overline{T}_i$	fixed period time nodes
$(\overline{T}_{i-1}, \overline{T}_i]$	i^{th} fixed side interval
$\overline{\Delta}_i$	$= \overline{T}_i - \overline{T}_{i-1}$ width of ith interval
κ_i	$= \kappa(\overline{T}_i)$ coupon over $(\overline{T}_i, \overline{T}_{i+1}]$
$\overline{\delta}_i$	$= \overline{\delta}(\overline{T}_i) = \overline{\Delta}_{i+1}$ coverage for κ_i
M	number fixed payment dates in a swap
$i0, ..., iM$	fixed side index in a swap
r	the regular roll $N = rM$
$j = J(i)$	floating $j \cong i$ on fixed side

A.2 Gaussian distributions

A.2.1 Conditional expectations

If the multi-dimensional vectors X and Y are jointly normally distributed, then (see [118]) X given Y is also normally distributed

$$(X \mid Y) \sim \mathbf{N}\left[\mathbf{E}(X \mid Y), \operatorname{var}(X \mid Y)\right]$$

where

$$\operatorname{cov}(X, Y) = \mathbf{E}(X - \mathbf{E}X)(Y - \mathbf{E}Y)^* \qquad \operatorname{var}(X) = \operatorname{cov}(X, X)$$

and

$$\begin{aligned} \mathbf{E}(X \mid Y) &= \mathbf{E}(X) + \operatorname{cov}(X, Y) \operatorname{var}^{-1}(Y)\left[Y - \mathbf{E}(Y)\right], \\ \operatorname{var}(X \mid Y) &= \operatorname{var}(X) - \operatorname{cov}(X, Y) \operatorname{var}^{-1}(Y) \operatorname{cov}(Y, X). \end{aligned}$$

A.2.2 Density shift

For a pair of jointly distributed normal random vectors X and Y with zero mean the following *density shift* formulae hold

$$\mathbf{E}\left[\exp\left(b^* X - \frac{1}{2}\operatorname{var} b^* X\right) f(Y)\right] = \mathbf{E} f\left(Y + \operatorname{cov}(Y, X) b\right), \tag{A.1}$$

and in particular if $Y = X \sim \mathbf{N}(0, \Delta)$

$$\mathbf{E}\exp\left(b^* X - \frac{1}{2}\operatorname{var} b^* X\right) f(X) = \mathbf{E} f(X + \Delta b). \tag{A.2}$$

PROOF The left hand side of (A.2) is

$$\begin{aligned} &\mathbf{E}\exp\left(b^* X - \frac{1}{2}\operatorname{var} b^* X\right) f(X) \\ &= \int_{\mathbb{R}^n} \exp\left(b^* x - \frac{1}{2} b^* \Delta b\right) f(x) \frac{1}{|\Delta|^{\frac{1}{2}} (2\pi)^{\frac{n}{2}}} \exp\left(-\frac{1}{2} x^* \Delta^{-1} x\right) dx. \end{aligned}$$

Remembering that Δ is symmetric, the exponential part simplifies like

$$-\frac{1}{2} x^* \Delta^{-1} x + b^* x - \frac{1}{2} b^* \Delta b = -\frac{1}{2}(x - \Delta b)^* \Delta^{-1}(x - \Delta b),$$

and so

$$\begin{aligned}
&\mathbf{E}\exp\left(b^*X - \frac{1}{2}\operatorname{var} b^*X\right) f(X) \\
&= \int_{\mathbb{R}^n} f(x) \frac{1}{|\Delta|^{\frac{1}{2}} (2\pi)^{\frac{n}{2}}} \exp\left(-\frac{1}{2}(x - \Delta b)^* \Delta^{-1}(x - \Delta b)\right) dx, \\
&= \int_{\mathbb{R}^n} f(x + \Delta b) \frac{1}{|\Delta|^{\frac{1}{2}} (2\pi)^{\frac{n}{2}}} \exp\left(-\frac{1}{2} x^* \Delta^{-1} x\right) dx, \\
&= \mathbf{E} f(X + \Delta b).
\end{aligned}$$

Conditioning on Y, the left hand side of (A.1) becomes

$$\begin{aligned}
&\mathbf{E}\left[\exp\left(b^*X - \frac{1}{2}\operatorname{var} b^*X\right) f(Y)\right] \\
&= \mathbf{E}\left[f(Y)\,\mathbf{E}\left\{\exp(b^*X\,|Y) - \frac{1}{2}\operatorname{var}(b^*X\,|Y)\right\}\right].
\end{aligned}$$

But from Section-, b^*X given Y is also normally distributed

$$(b^*X|\,Y) \sim \mathbf{N}\left[\mathbf{E}(b^*X\,|Y), \operatorname{var}(b^*X\,|Y)\right]$$

where

$$\begin{aligned}
\mathbf{E}(b^*X\,|Y) &= b^* \operatorname{cov}(X, Y) \operatorname{var}^{-1}(Y)\,Y, \\
\operatorname{var}(b^*X\,|Y) &= b^*\left[\operatorname{var}(X) - \operatorname{cov}(X, Y)\operatorname{var}^{-1}(Y)\operatorname{cov}(Y, X)\right] b, \\
\operatorname{cov}(b^*X, Y) &= \mathbf{E} b^* X Y^* = b^* \operatorname{cov}(X, Y),
\end{aligned}$$

so the inside conditional expectation simplifies to

$$\begin{aligned}
&\mathbf{E}\left\{\exp(b^*X\,|Y) - \frac{1}{2}\operatorname{var}(b^*X\,|Y)\right\} = \mathbf{E}\left\{\exp\left(b^*X - \frac{1}{2} b^* \operatorname{var}(X)\, b\right)|Y\right\}, \\
&= \exp\left\{b^* \operatorname{cov}(X, Y)\operatorname{var}^{-1}(Y)\,Y - \frac{1}{2} b^* \operatorname{cov}(X, Y)\operatorname{var}^{-1}(Y)\operatorname{cov}(Y, X)\, b\right\}, \\
&= \exp\left\{\left(b^* \operatorname{cov}(X, Y)\operatorname{var}^{-1}(Y)\,Y\right) - \frac{1}{2}\operatorname{var}\left(b^* \operatorname{cov}(X, Y)\operatorname{var}^{-1}(Y)\,Y\right)\right\}.
\end{aligned}$$

Hence, applying (A.2)

$$\begin{aligned}
&\mathbf{E}\left[\exp\left(b^*X - \frac{1}{2}\operatorname{var} b^*X\right) f(Y)\right] \\
&= \mathbf{E}\left[\exp\left\{\begin{array}{c} \left(b^* \operatorname{cov}(X, Y)\operatorname{var}^{-1}(Y)\,Y\right) \\ -\frac{1}{2}\operatorname{var}\left(b^* \operatorname{cov}(X, Y)\operatorname{var}^{-1}(Y)\,Y\right) \end{array}\right\} f(Y)\right], \\
&= \mathbf{E} f\left\{Y + \operatorname{var}(Y)\left[b^* \operatorname{cov}(X, Y)\operatorname{var}^{-1}(Y)\right]^*\right\}, \\
&= \mathbf{E} f(Y + \operatorname{cov}(Y, X)\, b).
\end{aligned}$$

$\square$

A.2.3 Black formula

Let X, Y be jointly normally distributed with zero mean and set $\zeta^2 = \operatorname{var}(X - Y)$. The *Black formula* is

$$\mathbf{B} = \mathbf{B}(K,\ L,\ \zeta) = \mathbf{E}\left[K \exp\left(X - \frac{1}{2}\operatorname{var} X\right) - L \exp\left(Y - \frac{1}{2}\operatorname{var} Y\right)\right]^{+},$$

$$= K\mathbf{N}(h) - L\mathbf{N}(h - \zeta) \quad \textit{where} \quad h = \frac{\ln\left(\frac{K}{L}\right) + \frac{1}{2}\zeta^2}{\zeta},$$

with its *Greeks* given by the partial derivatives

$$\frac{\partial \mathbf{B}}{\partial K} = \mathbf{N}(h), \quad \frac{\partial \mathbf{B}}{\partial L} = -\mathbf{N}(h - \zeta), \quad \frac{\partial \mathbf{B}}{\partial \zeta} = K\mathbf{N}'(h) = L\mathbf{N}'(h - \zeta),$$

$$\frac{\partial^2 \mathbf{B}}{\partial K^2} = \frac{1}{K\zeta}\mathbf{N}'(h), \quad \frac{\partial^2 \mathbf{B}}{\partial L^2} = \frac{1}{L\zeta}\mathbf{N}'(h - \zeta),$$

$$\frac{\partial^2 \mathbf{B}}{\partial K \partial L} = -\frac{1}{L\zeta}\mathbf{N}'(h) = -\frac{1}{K\zeta}\mathbf{N}'(h - \zeta).$$

PROOF Using (A.1)

$$\mathbf{E}\left[K \exp\left(X - \frac{1}{2}\operatorname{var} X\right) - L \exp\left(Y - \frac{1}{2}\operatorname{var} Y\right)\right]^{+}$$

$$= \mathbf{E}\exp\left(Y - \frac{1}{2}\operatorname{var} Y\right)\left[\begin{array}{c} K\exp\left\{(X - Y) - \frac{1}{2}\operatorname{var}(X - Y)\right\} \\ \times \exp\left(\operatorname{var} Y - \operatorname{cov}(X, Y)\right) \end{array} - L\right]^{+},$$

$$= \mathbf{E}\left[\begin{array}{c} K\exp\left\{(X - Y) - \frac{1}{2}\operatorname{var}(X - Y)\right\} \\ \times \exp\left(\operatorname{var} Y - \operatorname{cov}(X, Y) - \operatorname{var} Y + \operatorname{cov}(X, Y)\right) \end{array} - L\right]^{+},$$

$$= \mathbf{E}\left[K\exp\left\{(X - Y) - \frac{1}{2}\operatorname{var}(X - Y)\right\} - L\right]^{+},$$

and it is trivial to show this is equal to $\mathbf{B}(K,\ L,\ \zeta)$ as defined above.

The *Greeks* follow either by partially differentiating $\mathbf{B}(K,\ L,\ \zeta)$ directly, or in a similar fashion to the derivation of the Black formula by differentiating partially under the expectation and simplifying using Section-A.2.2 and Heaviside and Dirac functions, see Section-A.5. □

Thus at time-t a Black caplet or floorlet fixed at T, paid at T_1, struck at κ with implied volatility σ, have the same *zeta* ζ given by

$$\zeta^2 = \sigma^2 (T - t)$$

and their values are respectively

$$\operatorname{cpl}(t) = B(t, T_1)\,\mathbf{B}\left(K(t, T),\ \kappa,\ \sigma\sqrt{T - t}\right),$$

$$\operatorname{flt}(t) = B(t, T_1)\,\mathbf{B}\left(\kappa,\ K(t, T),\ \sigma\sqrt{T - t}\right).$$

A.2.4 Gaussian density derivatives

If $\mathbf{Z} : \mathbb{R}^m \times \mathbb{R}^{\frac{1}{2}m(m+1)} \mapsto \mathbb{R}$ is a *Gaussian density* in $\mathbb{R}^m$

$$\mathbf{Z}\{x, R\} = \frac{1}{|R|^{\frac{1}{2}} (2\pi)^{\frac{m}{2}}} \exp\left(-\frac{1}{2} x^* R^{-1} x\right),$$
$$x = (x_i) \quad R = (r_{i,j}) \quad r_{i,j} = r_{j,i},$$

then *derivatives of the density* satisfy

$$\frac{\partial \mathbf{Z}\{x, R\}}{\partial r_{i,i}} = \frac{1}{2} \frac{\partial^2 \mathbf{Z}\{x, R\}}{\partial x_i^2} \quad (i = j) \qquad \frac{\partial \mathbf{Z}\{x, R\}}{\partial r_{i,j}} = \frac{\partial^2 \mathbf{Z}\{x, R\}}{\partial x_i \partial x_j} \quad (i \neq j).$$

Note that we are not assuming $\mathbf{Z}\{x, R\}$ is a standard Gaussian density (though it can be), that is, we are not assuming the diagonal terms on R are 1s.

PROOF Introduce $c = (2\pi)^{-\frac{m}{2}}$, let $A = R^{-1}$ with $A = (a_{ij})$ be the corresponding covariance, and write

$$\mathbf{Z} = c\,|A|^{\frac{1}{2}} \exp\left(-\frac{1}{2} x^* A\, x\right).$$

For any i, j, on or off the diagonal, partial differentiation yields

$$\frac{1}{\mathbf{Z}} \frac{\partial^2 \mathbf{Z}}{\partial x_i \partial x_j} = \left\{ \frac{\partial\left(\frac{1}{2}x^* A\, x\right)}{\partial x_i} \frac{\partial\left(\frac{1}{2}x^* A\, x\right)}{\partial x_j} - \frac{\partial^2\left(\frac{1}{2}x^* A\, x\right)}{\partial x_i \partial x_j} \right\},$$
$$\frac{1}{\mathbf{Z}} \frac{\partial \mathbf{Z}}{\partial r_{i,j}} = \left\{ \frac{1}{|A|^{\frac{1}{2}}} \frac{\partial |A|^{\frac{1}{2}}}{\partial r_{i,j}} - \frac{\partial\left(\frac{1}{2}x^* A\, x\right)}{\partial r_{i,j}} \right\}.$$

The proof is completed by showing

$$-\frac{\partial^2\left(\frac{1}{2}x^* A\, x\right)}{\partial x_i \partial x_j} = \frac{1}{|A|^{\frac{1}{2}}} \frac{\partial |A|^{\frac{1}{2}}}{\partial r_{i,j}} \lambda_{i,j} \tag{A.3}$$
$$\frac{\partial\left(\frac{1}{2}x^* A\, x\right)}{\partial x_i} \frac{\partial\left(\frac{1}{2}x^* A\, x\right)}{\partial x_j} = -\frac{\partial\left(\frac{1}{2}x^* A\, x\right)}{\partial r_{i,j}} \lambda_{i,j}, \tag{A.4}$$
$$\textit{where} \quad \lambda_{i,i} = 2 \qquad \textit{and} \qquad \lambda_{i,j} = 1 \quad \textit{if} \quad i \neq j.$$

Clearly the left hand side of (A.3) is $-a_{i,j}$. Also, because $|A|^{\frac{1}{2}} = |R|^{-\frac{1}{2}}$

$$\frac{\partial |A|^{\frac{1}{2}}}{\partial r_{i,j}} = -\frac{1}{2} |R|^{-\frac{3}{2}} \frac{\partial |R|}{\partial r_{i,j}} = -\frac{1}{2} |A|^{\frac{1}{2}} \frac{1}{|R|} \frac{\partial |R|}{\partial r_{i,j}}.$$

But, if $R_{i,j}$ is the signed conjugate of $r_{i,j}$ in R

$$\frac{1}{|R|} \frac{\partial |R|}{\partial r_{i,j}} = \begin{cases} 2\frac{R_{i,j}}{|R|} = 2a_{i,j} & \textit{if} \quad i \neq j \\ \frac{R_{i,i}}{|R|} = a_{i,i} & \textit{if} \quad i = j \end{cases},$$

and (A.3) follows.

The left hand side of (A.4) is just the ith row of Ax multiplied by the jth row of Ax. On the other side, start with $AR = \mathbf{I}_m$, differentiate, multiply by A, and get

$$\frac{\partial A}{\partial r_{i,j}} = -A\frac{\partial R}{\partial r_{i,j}}A = -A\,\mathbf{I}_{i,j}\,A,$$

where $\mathbf{I}_{i,j}$ is the $m \times m$ matrix with 1 in the (i,j) and (j,i) positions and 0 elsewhere. Hence

$$-\frac{\partial\left(\frac{1}{2}x^* A\,x\right)}{\partial r_{i,j}} = \frac{1}{2}x^* A\,\mathbf{I}_{i,j}\,Ax = \frac{1}{2}(xA)^* \begin{pmatrix} 0 \\ row\ i\ contains\ j^{th}\ row\ Ax \\ 0 \\ row\ j\ contains\ i^{th}\ row\ Ax \\ 0 \end{pmatrix},$$

$$= \begin{cases} (Ax)_i\,(Ax)_j & if \quad i \neq j, \\ \frac{1}{2}(Ax)_i^2 & if \quad i = j. \end{cases},$$

and (A.4) follows. □

This result has a lot of unexpected applications. For example, in the next Section-A.2.5 it gives a connection between the gammas and vegas of a multi-asset European option.

Another example is a power series expansion for the bivariate Gaussian density in terms of univariate densities

$$\mathbf{Z}_2(x,y;\rho) = \sum_{n=0}^{\infty} \frac{\mathbf{Z}^{(n)}(x)\,\mathbf{Z}^{(n)}(y)}{n!}\rho^n. \tag{A.5}$$

Repeated application of the result to the bivariate density $\mathbf{Z}_2(x,y;\rho)$ with $x_1 = x$, $x_2 = y$ and $r_{1,2} = \rho$ gives

$$\frac{\partial^n}{\partial \rho^n}\mathbf{Z}_2(x,y;\rho) = \frac{\partial^{2n}}{\partial x^n \partial y^n}\mathbf{Z}_2(x,y;\rho),$$

from which it follows, setting $\rho = 0$, that

$$\left.\frac{\partial^n}{\partial \rho^n}\mathbf{Z}_2(x,y;\rho)\right|_{\rho=0} = \left.\frac{\partial^{2n}}{\partial x^n \partial y^n}\mathbf{Z}_2(x,y;\rho)\right|_{\rho=0} = \mathbf{Z}^{(n)}(x)\,\mathbf{Z}^{(n)}(y).$$

A standard Taylors series expansion of $\mathbf{Z}_2(x,y;\rho)$ in terms of ρ, then yields (A.5). Note that this result can be easily generalized to multi-variate Gaussian densities.

A.2.5 Gamma and vega connection

For a European option with payoff $g : \mathbb{R}^m \mapsto \mathbb{R}$ and present value

$$C = \mathbf{E}g\left[S_1 \exp\left(X_1 - \frac{1}{2}\operatorname{var} X_1\right), .., S_m \exp\left(X_m - \frac{1}{2}\operatorname{var} X_m\right)\right], \quad \text{(A.6)}$$

$$\textit{where} \qquad X = (X_i) \sim \mathbf{N}(0, R), \quad R = (r_{i,j}),$$

$$\textit{if} \qquad \Lambda = (\Lambda_{i,j}) = \left(\frac{\partial C}{\partial r_{i,j}}\right) \qquad \Gamma = (\Gamma_{i,j}) = \left(\frac{\partial^2 C}{\partial S_i \partial S_j}\right)$$

$$\textit{then} \qquad \Lambda_{i,i} = \frac{1}{2}S_i^2\,\Gamma_{i,i}\ (i = j) \qquad \Lambda_{i,j} = S_i\,S_j\,\Gamma_{i,j}\ (i \neq j).$$

Thus, for example, a standard Black-Scholes option struck at K, exercising at T, on a stock with present value S and volatility σ in the presence of constant interest rates r, would be given by

$$C = \mathbf{E}\left[S \exp\left(X - \frac{1}{2}\operatorname{var} X\right) - K \exp(-rT)\right]^+,$$

$$\textit{where} \qquad X \sim \mathbf{N}(0, R), \quad R = \sigma^2 T.$$

So from Section-A.2.3 setting $\zeta^2 = R$, $K = S$ and $L = K\exp(-rT)$

$$\frac{\partial \mathbf{B}}{\partial \zeta} = K\mathbf{N}'(h), \quad \Lambda = \frac{\partial C}{\partial R} = S\mathbf{N}'(h), \quad \Gamma = \frac{\partial^2 \mathbf{B}}{\partial K^2} = \frac{1}{S\ R}\mathbf{N}'(h).$$

Note that while our gammas are quite standard, our vegas are slightly different from the conventional ones in that they include off diagonal covariance terms, they include the time dependency or theta, and there are as many individual vegas $\Lambda_{i,j}$ as gammas $\Gamma_{i,i}$.

For an option over a single index it is quite easy to prove that $\Lambda = \frac{1}{2}S^2\Gamma$, and for multiple indices this generalizes to

Case $i \neq j$. Take $\mathbf{e}(x_i)$ to mean

$$\mathbf{e}(x_i) = \exp\left(x_i - \frac{1}{2}r_{i,i}\right),$$

and note that the $r_{i,i}$ do not enter calculations in this case. Starting from the definition (A.6), integrating twice by parts with respect to x_i and x_j, and using Section-A.2.4, we get

$$\begin{aligned}
\Gamma_{i,j} &= \int_{\mathbb{R}^m} g^{(i,j)}\left[S_1\mathbf{e}(x_1), S_2\mathbf{e}(x_2), .., S_m\mathbf{e}(x_m)\right]\mathbf{e}(x_i)\,\mathbf{e}(x_j)\,\mathbf{N}(x, R)\,dx, \\
&= \int_{\mathbb{R}^m} \frac{1}{S_i\,S_j} g\left[S_1\mathbf{e}(x_1), S_2\mathbf{e}(x_2), .., S_m\mathbf{e}(x_m)\right]\mathbf{N}^{(i,j)}(x, R)\,dx, \\
&= \frac{1}{S_i\,S_j}\int_{\mathbb{R}^m} g\left[S_1\mathbf{e}(x_1), S_2\mathbf{e}(x_2), .., S_m\mathbf{e}(x_m)\right]\frac{\partial}{\partial r_{i,j}}\mathbf{N}(x, R)\,dx, \\
&= \frac{1}{S_i\,S_j}\Lambda_{i,j}.
\end{aligned}$$

Case $i = j$. In this case the $r_{i,i}$ in the $\mathbf{e}(x_i)$ must be considered. Integrating twice by parts with respect to x_i, the gamma is

$$\begin{aligned}\Gamma_{i,i} &= \int_{\mathbb{R}^m} g^{(i,i)}\left[S_1\mathbf{e}(x_1), S_2\mathbf{e}(x_2), .., S_m\mathbf{e}(x_m)\right]\mathbf{e}^2(x_i)\,\mathbf{N}(x,R)\,dx,\\ &= \int_{\mathbb{R}^m} \frac{1}{S_i^2} g\left[S_1\mathbf{e}(x_1), S_2\mathbf{e}(x_2), .., S_m\mathbf{e}(x_m)\right]\mathbf{N}^{(i,i)}(x,R)\,dx\\ &\quad - \int_{\mathbb{R}^m} \frac{1}{S_i} g^{(i)}\left[S_1\mathbf{e}(x_1), S_2\mathbf{e}(x_2), .., S_m\mathbf{e}(x_m)\right]\mathbf{e}(x_i)\,\mathbf{N}(x,R)\,dx,\end{aligned}$$

while the vega is

$$\begin{aligned}\Lambda_{i,i} &= \int_{\mathbb{R}^m} g\left[S_1\mathbf{e}(x_1), S_2\mathbf{e}(x_2), .., S_m\mathbf{e}(x_m)\right]\frac{\partial}{\partial r_{i,i}}\mathbf{N}(x,R)\,dx\\ &\quad - \int_{\mathbb{R}^m} g^{(i)}\left[S_1\mathbf{e}(x_1), S_2\mathbf{e}(x_2), .., S_m\mathbf{e}(x_m)\right]\frac{1}{2}S_i\mathbf{e}(x_i)\,\mathbf{N}(x,R)\,dx,\\ &= \frac{1}{2}S_i^2\,\Gamma_{i,i}.\end{aligned}$$

A.2.6 Bivariate distribution

From Abramowitz [1], the *bivariate normal* distribution is defined by

$$\begin{pmatrix}X\\Y\end{pmatrix} \sim \mathbf{N}\left(\begin{pmatrix}0\\0\end{pmatrix}, \begin{pmatrix}1 & \rho\\ \rho & 1\end{pmatrix}\right),$$

$$\mathbf{N}_2(a,b;\rho) = \mathbf{E}\mathbb{I}\{X \le a\}\mathbb{I}\{Y \le b\} = \int_{-\infty}^{a}\int_{-\infty}^{b} Z_2(x,y;\rho)\,dxdy,$$

$$Z_2(x,y;\rho) = \frac{1}{2\pi\sqrt{1-\rho^2}}\exp\left\{-\frac{1}{2(1-\rho^2)}\left[x^2 - 2\rho xy + y^2\right]\right\}.$$

Using the properties of conditional expectations

$$\begin{aligned}\mathbf{N}_2(a,b;\rho) &= \mathbf{E}\mathbb{I}(X \le a)\,\mathbb{I}(Y \le b) = \mathbf{E}\left\{\mathbb{I}(X \le a)\,\mathbf{E}\left[\mathbb{I}(Y \le b)|\,X\right]\right\},\\ &= \mathbf{E}\left\{\mathbb{I}(X \le a)\,\mathbf{N}\left(\frac{b-\rho X}{\sqrt{1-\rho^2}}\right)\right\} = \int_{-\infty}^{a}\mathbf{N}\left(\frac{b-\rho x}{\sqrt{1-\rho^2}}\right)\mathbf{N}'(x)\,dx.\end{aligned}$$

The *bivariate generator* L_2 is given by

$$dX(t) = \rho dW_1(t), \quad dY(t) = \rho dW_1(t) + \sqrt{1-\rho^2}dW_2(t),$$

$$L_2 = \frac{1}{2}\left(\frac{\partial^2}{\partial x^2} + \frac{\partial^2}{\partial y^2}\right) + \rho\frac{\partial^2}{\partial x\partial y}.$$

A.2.7 Ratio of cumulative and density distributions

This result is essential for inverting normal distributions in the tails; for example, when finding implied volatilities of short-dated options away from the money. Setting

$$F(x) = x + \cfrac{1}{x + \cfrac{2}{x + \cfrac{3}{x + \cfrac{4}{x + \dots}}}},$$

then, see Abramowitz [1],

$$\frac{\mathbf{N}_1(x)}{1 - \mathbf{N}(x)} = F(x) \quad x > 0 \qquad \text{and} \qquad \frac{\mathbf{N}_1(x)}{\mathbf{N}(x)} = F(-x) \quad x < 0.$$

A.2.8 Expected values of normals

Let $\mathbf{N}(\cdot)$ denote the standard normal cumulative distribution function, and introduce

$$\mathbf{N}'(x) = \mathbf{N}^{(1)}(x) = \frac{1}{\sqrt{2\pi}} \exp\left(-\frac{1}{2}x^2\right), \qquad \mathbf{N}^{(n)}(x) = \frac{d^n}{dx^n}\mathbf{N}(x),$$

$$\mathbf{\Phi}(x) = \mathbf{N}^{(-1)}(x) = \int_{-\infty}^{x} \mathbf{N}(u)\, du.$$

Note that

$$\mathbf{N}^{(2)}(x) = -x\frac{1}{\sqrt{2\pi}} \exp\left(-\frac{1}{2}x^2\right) = -x\mathbf{N}^{(1)}(x) \quad \textit{and}$$

$$\begin{aligned}\mathbf{\Phi}(x) &= \int_{-\infty}^{x} \mathbf{N}(u)\, du = x\mathbf{N}(x) - \int_{-\infty}^{x} u\mathbf{N}^{(1)}(u)\, du,\\ &= x\mathbf{N}(x) + \int_{-\infty}^{x} \mathbf{N}^{(2)}(u)\, du = \mathbf{N}'(x) + x\mathbf{N}(x).\end{aligned}$$

Combinations and adaptions of the following formulae are often useful. If $X \sim \mathbf{N}(0,1)$ then

$$\begin{aligned}\mathbf{E}\,\mathbf{N}(aX+b) &= \mathbf{N}\left(\frac{b}{\sqrt{a^2+1}}\right),\\ \mathbf{E}\,\mathbf{N}^{(n)}(aX+b) &= \left(\frac{1}{\sqrt{a^2+1}}\right)^n \mathbf{N}^{(n)}\left(\frac{b}{\sqrt{a^2+1}}\right),\\ \mathbf{E}X\,\mathbf{N}(aX+b) &= a\mathbf{E}\mathbf{N}^{(1)}(aX+b) = \frac{a}{\sqrt{a^2+1}}\mathbf{N}^{(1)}\left(\frac{b}{\sqrt{a^2+1}}\right),\\ \mathbf{E}\,\mathbf{\Phi}(aX+b) &= \sqrt{a^2+1}\mathbf{\Phi}\left(\frac{b}{\sqrt{a^2+1}}\right).\end{aligned}$$

PROOF To establish the *first* formula partially differentiate

$$\mathbf{I}(a,b) = \mathbf{E}\mathbf{N}(aX+b)$$

with respect to b, simplify

$$\begin{aligned}
\partial_b \mathbf{I}(a,b) &= \int_{-\infty}^{\infty} \mathbf{N}'(ax+b)\,\mathbf{N}'(x)\,dx \\
&= \frac{1}{2\pi}\int_{-\infty}^{\infty} \exp\left[-\frac{1}{2}\left\{(ax+b)^2+x^2\right\}\right]dx, \\
= \frac{1}{2\pi}\exp\left[-\frac{1}{2}\frac{b^2}{a^2+1}\right] &\int_{-\infty}^{\infty} \exp\left[-\frac{1}{2}\left\{\sqrt{(a^2+1)}x+\frac{ab}{\sqrt{(a^2+1)}}\right\}^2\right]dx, \\
&= \frac{1}{\sqrt{2\pi\left(a^2+1\right)}}\exp\left[-\frac{1}{2}\frac{b^2}{a^2+1}\right],
\end{aligned}$$

and integrate back to get

$$\begin{aligned}
\mathbf{I}(a,b) &= \int_{-\infty}^{b} \frac{1}{\sqrt{2\pi\left(a^2+1\right)}}\exp\left[-\frac{1}{2}\frac{u^2}{\left(a^2+1\right)}\right]du + \text{function}(a) \\
&= \mathbf{N}\left(\frac{b}{\sqrt{a^2+1}}\right), \quad \textit{because} \quad \mathbf{I}(a,0)=\frac{1}{2} \quad \Rightarrow \quad \text{function}(a)=0.
\end{aligned}$$

The *second* formula follows by partially differentiating the *first* formula n times with respect to b.

For the *third* formula

$$\begin{aligned}
&\mathbf{E}X\,\mathbf{N}(aX+b) \\
&= \int_{-\infty}^{\infty} \mathbf{N}(ax+b)\;x\mathbf{N}^{(1)}(x)\,dx = -\int_{-\infty}^{\infty} \mathbf{N}(ax+b)\;\mathbf{N}^{(2)}(x)\,dx, \\
&= \int_{-\infty}^{\infty} a\,\mathbf{N}^{(1)}(ax+b)\;\mathbf{N}^{(1)}(x)\,dx = a\mathbf{E}\mathbf{N}^{(1)}(aX+b).
\end{aligned}$$

The final formula requires the second and third

$$\begin{aligned}
&\mathbf{E}\boldsymbol{\Phi}(aX+b) \\
&= \mathbf{E}\left\{\mathbf{N}^{(1)}(aX+b)+aX\mathbf{N}(aX+b)+b\mathbf{N}(aX+b)\right\}, \\
&= \sqrt{a^2+1}\left\{\mathbf{N}^{(1)}\left(\frac{b}{\sqrt{a^2+1}}\right)+\frac{b}{\sqrt{a^2+1}}\mathbf{N}\left(\frac{b}{\sqrt{a^2+1}}\right)\right\}.
\end{aligned}$$

□

A.3 Stochastic calculus

A.3.1 Multi-dimensional Ito

If $f : \mathbb{R}^{+} \times \mathbb{R}^{n} \to \mathbb{R}$ has a continuous first order partial derivative in $t \in \mathbb{R}^{+}$ and continuous second order partial derivatives in $x \in \mathbb{R}^{n}$, and X_t is an n-dimensional Ito process, then

$$df(t, X_t) = \frac{\partial}{\partial t} f(t, X_t)\, dt + [\nabla f(t, X_t)]^{*}\, dX_t + \frac{1}{2} \operatorname{trace} \nabla\nabla^{*} f(t, X_t)\, d\langle X\rangle_t$$

where ∇ is the gradient operator, and $\langle X\rangle_t$ is the quadratic variation matrix

$$\langle X\rangle_t = \left(\left\langle X^{(i)}, X^{(j)}\right\rangle_t\right), \quad i, j = 1, ..., n.$$

A.3.2 Brownian bridge

Let M_t be a continuous Gaussian martingale with absolutely continuous quadratic variation q_t

$$M_t = \int_0^t \gamma(u)\, dW_u, \qquad q_t = \int_0^t \gamma^2(u)\, du.$$

If $0 < s < t$, then $(M_s | M_t)$ is normally distributed

$$(M_s | M_t) \sim \mathbf{N}\left(\frac{q_s}{q_t} M_t, \quad \frac{q_s}{q_t}[q_t - q_s]\right),$$

and the conditional expectation

$$\mathbf{E}\{\mathcal{E}(M_s) | M_t\} = \exp\left\{\frac{q_s}{q_t} M_t - \frac{1}{2}\frac{q_s^2}{q_t}\right\}.$$

A.3.3 Product and quotient processes

If X_t and Y_t are positive Ito processes with SDEs

$$\frac{dX_t}{X_t} = \mu_X dt + \sigma_X^{*} dW_t, \quad \frac{dY_t}{Y_t} = \mu_Y dt + \sigma_Y^{*} dW_t,$$

their product and quotient processes are also Ito processes with SDEs

$$\frac{d(XY)_t}{(XY)_t} = (\mu_X + \mu_Y + \sigma_X^{*} \sigma_Y)\, dt + (\sigma_X + \sigma_Y)^{*}\, dW_t,$$

$$\frac{d\left(\frac{X}{Y}\right)_t}{\left(\frac{X}{Y}\right)_t} = [\mu_X - \mu_Y - \sigma_Y^{*}(\sigma_X - \sigma_Y)]\, dt + (\sigma_X - \sigma_Y)^{*}\, dW_t.$$

PROOF For the product process, we have

$$d(XY)_t = Y_t dX_t + X_t dY_t + d\langle X, Y\rangle_t,$$

so dividing by X_tY_t the result follows from

$$\frac{d(XY)_t}{(XY)_t} = \frac{dX_t}{X_t} + \frac{dY_t}{Y_t} + \frac{d\langle X, Y\rangle_t}{X_tY_t} = \mu_X dt + \sigma_X^* dW_t + \mu_Y dt + \sigma_Y^* dW_t + \sigma_X^* \sigma_Y dt.$$

Applying Ito

$$d\left(\frac{1}{Y_t}\right) = -\frac{1}{Y_t^2} dY_t + \frac{1}{Y_t^3} d\langle Y\rangle_t, \qquad \Rightarrow$$

$$\frac{d\left(\frac{1}{Y_t}\right)}{\frac{1}{Y_t}} = -\frac{dY_t}{Y_t} + \frac{d\langle Y\rangle_t}{Y_t^2} = \left(-\mu_Y + \sigma_Y^2\right) dt - \sigma_Y^* dW_t.$$

on dividing by $\frac{1}{Y_t}$. The quotient result follows on writing

$$\left(\frac{X}{Y}\right)_t = X_t \frac{1}{Y_t}$$

and using the product process result. ∎

A.3.4 Conditional change of measure

Let $\mathbb{P}$ and $\mathbb{Q}$ be equivalent measures with respect to a σ-algebra $\mathcal{F}$ ($\mathbb{Q} = Z\mathbb{P}$ or $\mathbf{E}_{\mathbb{Q}}(Y) = \mathbf{E}_{\mathbb{P}}(ZY)$), and suppose Y is $\mathcal{F}$-measurable. If $\mathcal{G}$ is a sub σ-algebra of $\mathcal{F}$. ($\mathcal{G} \subset \mathcal{F}$), then

$$\mathbf{E}_{\mathbb{Q}}\{Y|\mathcal{G}\} = \frac{\mathbf{E}_{\mathbb{P}}\{ZY|\mathcal{G}\}}{\mathbf{E}_{\mathbb{P}}\{Z|\mathcal{G}\}}. \tag{A.7}$$

PROOF The right-hand side of (A.7) is $\mathcal{G}$-measurable, and so for any $A \in \mathcal{G}$

$$\begin{aligned}\mathbf{E}_{\mathbb{Q}}\left\{\mathbb{I}(A)\frac{\mathbf{E}_{\mathbb{P}}\{ZY|\mathcal{G}\}}{\mathbf{E}_{\mathbb{P}}\{Z|\mathcal{G}\}}\right\} &= \mathbf{E}_{\mathbb{P}}\left\{Z\,\mathbb{I}(A)\frac{\mathbf{E}_{\mathbb{P}}\{ZY|\mathcal{G}\}}{\mathbf{E}_{\mathbb{P}}\{Z|\mathcal{G}\}}\right\} \\ &= \mathbf{E}_{\mathbb{P}}\{\mathbb{I}(A)\,\mathbf{E}_{\mathbb{P}}(ZY|\mathcal{G})\} = \mathbf{E}_{\mathbb{P}}\{Z\,\mathbb{I}(A)\,Y\} \\ &= \mathbf{E}_{\mathbb{Q}}\{\mathbb{I}(A)\,Y\} = \mathbf{E}_{\mathbb{Q}}\{\mathbb{I}(A)\,\mathbf{E}_{\mathbb{Q}}\{Y|\mathcal{G}\}\}.\end{aligned}$$

∎

A.3.5 Girsanov theorem

If $\frac{d\mathbb{Q}}{d\mathbb{P}} = Z_T$ and $Z_t = \mathbf{E}_{\mathbb{P}}\{Z_T|\mathcal{F}_t\} = \mathcal{E}\{Y_t\}$ (so $dZ_t = Z_t dY_t$, $\mathbf{E}_{\mathbb{P}}Z_t = 1$), then M_t is a $\mathbb{P}$-martingale iff $M_t - \langle M, Y\rangle_t$ is a $\mathbb{Q}$-martingale.

PROOF Setting $\widetilde{M}_t = M_t - \langle M, Y\rangle_t$, from (A.7) and using Ito's lemma

$$
\begin{aligned}
\mathbf{E}_{\mathbb{Q}}\left\{\widetilde{M}_t \middle| \mathcal{F}_s\right\} &= \frac{\mathbf{E}_{\mathbb{P}}\left\{ Z_t\left[M_t - \langle M,Y\rangle_t\right] \middle| \mathcal{F}_s\right\}}{\mathbf{E}_{\mathbb{P}}\left\{ Z_t \middle| \mathcal{F}_s\right\}} \\
&= \frac{1}{Z_s}\mathbf{E}_{\mathbb{P}}\left\{ Z_t\left[M_t - \langle M,Y\rangle_t\right] \middle| \mathcal{F}_s\right\}, \\
&= \frac{1}{Z_s}\mathbf{E}_{\mathbb{P}}\left\{ \begin{array}{c} \int_0^t Z_u\left[dM_u - d\langle M,Y\rangle_u\right] \\ + \int_0^t \left[M_u - \langle M,Y\rangle_u\right] dZ_u + \langle Z,M\rangle_t \end{array} \middle| \mathcal{F}_s \right\}, \\
&= \frac{1}{Z_s}\mathbf{E}_{\mathbb{P}}\left\{ \int_0^t Z_u dM_u + \int_0^t \left[M_u - \langle M,Y\rangle_u\right] dZ_u \middle| \mathcal{F}_s \right\}, \\
&= \frac{1}{Z_s}\left[\int_0^s Z_u dM_u + \int_0^s \left[M_u - \langle M,Y\rangle_u\right] dZ_u \right],
\end{aligned}
$$

because both Z_t and M_t are $\mathbb{P}_0$-martingales, and therefore so are stochastic integrals with respect to them. Hence

$$
\begin{aligned}
\mathbf{E}_{\mathbb{Q}}\left\{\widetilde{M}_t \middle| \mathcal{F}_s\right\} &= \frac{1}{Z_s}\left[\begin{array}{c} \int_0^s Z_u\left[dM_u - d\langle M,Y\rangle_u\right] \\ + \int_0^s \left[M_u - \langle M,Y\rangle_u\right] dZ_u + \langle Z,M\rangle_s \end{array} \right] \\
&= \frac{1}{Z_s} Z_s\left[M_s - \langle M,Y\rangle_s\right] = \widetilde{M}_s,
\end{aligned}
$$

showing $\widetilde{M}_t$ is a $\mathbb{Q}$-martingale. A similar argument establishes necessity. □

If W_t is $\mathbb{P}$-Brownian motion and $\mathbb{Q} = \mathcal{E}\left\{\int_0^T H_s dW_s\right\}\mathbb{P}$ so $Y_t = \int_0^t H_s dW_s$, then $\mathbb{Q}$-Brownian motion is $\widetilde{W}_t$ where

$$
\widetilde{W}_t = W_t - \left\langle W, \int_0^{\cdot} H_s dW_s \right\rangle_t \qquad or \qquad d\widetilde{W}_t = dW_t - H_t dt,
$$

and we can talk about the *change of measure* defined by $d\widetilde{W}_t = dW_t - H_t dt$.

Also, if W_t is n-component Brownian motion, with correlation between the i and j components given by $\rho_{i,j}\ dt = d\left\langle W_t^{(i)}, W_t^{(j)}\right\rangle$, and

$$
Y_t = \int_0^t H_s dW_s = \int_0^t \sum_{i=1}^n H_s^{(i)} dW_s^{(i)},
$$

then the correlation $\rho_{i,j}$ will appear in the expression for $\widetilde{W}_t$

$$
d\widetilde{W}_t^{(i)}(t) = dW_t^{(i)}(t) + \sum_{j=1}^n \rho_{i,j} H_t^{(j)}\ dt \quad (i = 1, ..., n).
$$

A.3.6 One-dimensional Ornstein Uhlenbeck process

This is the one and only one-dimensional stationary Gauss Markov process.

Constants	$A(t) = -\lambda \quad a(t) = 0 \quad \sigma(t) = \sigma$
Initial values	$X_0 = x \quad m(0) = x \quad V(0) = 0$
SDE	$dX_t = -\lambda X_t + \sigma dW_t$
Solution	$X_t = e^{-\lambda t}\left(x + \int_0^t \sigma e^{\lambda s} dW_s\right)$
Mean	$m(t) = e^{-\lambda t} x$
Autocorrelation	$\rho(s,t) = \frac{\sigma^2}{2\lambda} e^{-\lambda t}\left(e^{\lambda s} - e^{-\lambda s}\right)$
Variance	$V(t) = \frac{\sigma^2}{2\lambda}\left(1 - e^{-2\lambda t}\right)$
Conditional expectation	$U_s = \mathbf{E}(X_t \mid X_s) = e^{-\lambda(t-s)} X_s$
Conditional variance	$\mathrm{Var}(X_t \mid X_s) = \frac{\sigma^2}{2\lambda}\left(1 - e^{-2\lambda(t-s)}\right)$

The best way to estimate λ is by linear regression of ΔX_t against X_t.

A.3.7 Generalized multi-dimensional OU process

The multi-dimensional Ornstein Uhlenbeck process is the one and only Gauss Markov process.

OU SDE	$dX_t = [A(t) X_t + a(t)] dt + \sigma(t) dW_t, \quad 0 \le t < \infty$
Introduce	$\Phi'(t) = A(t)\Phi(t), \quad \Phi(0) = \mathbf{I}_n$
Solution	$X_t = \Phi(t)\left[X_0 + \int_0^t \Phi^{-1}(s) a(s) ds + \int_0^t \Phi^{-1}(s)\sigma(s) dW_s\right]$
Mean	$m(t) = \mathbf{E}X_t = \Phi(t)\left[m(0) + \int_0^t \Phi^{-1}(s) a(s) ds\right]$
	$m'(t) = A(t) m(t) + a(t)$
Autocorr	$\rho(s,t) = \mathbf{E}\{X_s - m(s)\}\{X_t - m(t)\}^*$
	$= \Phi(s)\left[\rho(0,0) + \int_0^{s\wedge t} \Phi^{-1}(u)\sigma(u)\sigma^*(u)\Phi^{-1^*}(u) du\right]\Phi^*(t)$
Variance	$V(t) = \rho(t,t)$
	$V'(t) = A(t)V(t) + V(t)A^*(t) + \sigma(t)\sigma^*(t)$
Dist	$X_t \sim \mathbf{N}\{m(t), V(t)\}$

Again the best way to estimate a constant mean reversion parameter $A(t) = A$, is by multi-linear regression of ΔX_t against X_t.

A.3.8 SDE of a discounted variable

Let $f(t, X_t)$ be an arbitrage free instrument in which the driver X_t is a diffusion with SDE

$$dX_t = \mu(t, X_t) dt + \sigma(t, X_t) dW_0(t)$$

under the spot measure $\mathbb{P}_0$. Because $\beta(t) = \exp\left(\int_0^t r(s)\,ds\right)$ is a finite variation process, $\frac{f(t,X_t)}{\beta(t)}$ is a $\mathbb{P}_0$ martingale, and stochastic and drift parts on both sides of an equation must match, that is

$$d\left\{\frac{f(t,X_t)}{\beta(t)}\right\} = (\text{expression})\,dW_0(t) = \frac{1}{\beta(t)}\frac{\partial}{\partial x}f(t,X_t)\;\sigma(t,X_t)\,dW_0(t)$$

$$\Rightarrow \quad df(t,X_t) = f(t,X_t)\,r(t)\,dt + \frac{\partial}{\partial x}f(t,X_t)\;\sigma(t,X_t)\,dW_0(t).$$

A.3.9 Ito-Venttsel formula

Let W_t be multi-dimensional Brownian motion, and suppose $F(t,u)$ is twice differentiable with respect to the parameter u and has an SDE of form

$$dF(t,u) = A(t,u)\,dt + B^*(t,u)\,dW_t.$$

If u_t satisfies the SDE

$$du_t = C(t,u_t)\,dt + D^*(t,u_t)\,dW_t,$$

then (see [127] and [110]) an SDE for $F(t,u_t)$ is

$$\begin{aligned} dF(t,u_t) &= A(t,u_t)\,dt + B^*(t,u_t)\,dW_t \\ &\quad + \frac{\partial}{\partial u}F(t,u_t)\,du_t + \frac{1}{2}\frac{\partial^2}{\partial u^2}F(t,u_t)\,|D(t,u_t)|^2\,dt \\ &\quad + \frac{\partial}{\partial u}B^*(t,u_t)\;D(t,u_t)\,dt. \end{aligned}$$

A.4 Linear Algebra

A.4.1 Cholesky decomposition

This is a convenient way of constructing a set of normally distributed random variables with given variances and correlations, and is very useful for simulation. We give the 3-dimensional version; simply truncate to the first two components for two dimensions.

Let $U \sim \mathbf{N}(0,\mathbf{I}_3)$, then the vector $X = \Gamma U$, where

$$\Gamma = \begin{pmatrix} \sigma_1 & 0 & 0 \\ \sigma_2\rho_{1,2} & \sigma_2\sqrt{1-\rho_{1,2}^2} & 0 \\ \sigma_3\rho_{1,3} & \sigma_3\dfrac{\left(\rho_{2,3}-\rho_{1,3}\rho_{1,2}\right)}{\sqrt{1-\rho_{1,2}^2}} & \sigma_3\dfrac{\sqrt{1-\rho_{1,3}^2-\rho_{1,2}^2-\rho_{2,3}^2+2\rho_{2,3}\rho_{1,3}\rho_{1,2}}}{\sqrt{\left(1-\rho_{1,2}^2\right)}} \end{pmatrix},$$

is distributed $\mathbf{N}(0, \Delta)$, with covariance matrix

$$\Delta = \Gamma\Gamma^* = \begin{pmatrix} \sigma_1 & 0 & 0 \\ 0 & \sigma_2 & 0 \\ 0 & 0 & \sigma_3 \end{pmatrix} \begin{pmatrix} 1 & \rho_{1,2} & \rho_{1,3} \\ \rho_{1,2} & 1 & \rho_{2,3} \\ \rho_{1,3} & \rho_{2,3} & 1 \end{pmatrix} \begin{pmatrix} \sigma_1 & 0 & 0 \\ 0 & \sigma_2 & 0 \\ 0 & 0 & \sigma_3 \end{pmatrix}.$$

A.4.2 Singular value decomposition

A square $n \times n$ matrix V is *orthogonal* if $V^T V = I_n$; that is, its columns are pairwise orthogonal and form an orthogonal basis for $\mathbb{R}^n$. It follows that its rows are also pairwise orthogonal $VV^T = I_n$ and form a basis for $\mathbb{R}^n$, because

$$V^T V = I_n \Rightarrow V^T V V^T = V^T \Rightarrow V^T \left(V V^T - I_n \right) = 0 \Rightarrow V V^T - I_n = 0.$$

Let A be any real $m \times n$ matrix; that is, any one of $m < n$ or $m = n$ or $m > n$ is permissible. If $Av = wu$, then w is called a *singular value* of A with corresponding *singular vectors* v and u. The *singular value decomposition* (SVD) algorithm produces the U, V and W matrices in the following theorem (see either [47], or [98] with a slightly different notation for U and W).

THEOREM A.1
If A is a real $m \times n$ matrix, there exists orthogonal matrices

$$U = [u_1, u_2, ..., u_m] \in \mathbb{R}^{m \times m}, \qquad V = [v_1, v_2, ..., v_n] \in \mathbb{R}^{n \times n}$$

and a diagonal matrix W such that

$$A = U\ W\ V^T, \qquad W = \operatorname{diag}(w_1, w_2, ..., w_p) \in \mathbb{R}^{m \times n}, \quad p = \min\{m, n\},$$

where $w_1 \geq w_2 \geq ..... \geq w_p \geq 0$.

Some very practical uses for the SVD stem from the next theorem.

THEOREM A.2
Suppose $A = U\ W\ V^T$ is the SVD of $A \in \mathbb{R}^{m \times n}$ with $r = \operatorname{rank} A$. If $b \in \mathbb{R}^m$, then

$$\widetilde{x} = \sum_{j=1}^{r} \frac{u_j^T b}{w_j} v_j \tag{A.8}$$

minimizes $\|Ax - b\|_2$ and has the smallest 2-norm of all minimizers. Moreover

$$\|Ax - b\|_2^2 = \sum_{j=r+1}^{m} \left(u_j^T b \right)^2.$$

If A is not rank deficient, that is, $r = \operatorname{rank} A = \min\{m, n\}$ (interestingly the SVD can identify and remedy such deficiencies, see [47]), then all the w_j are strictly positive and we can form an $n \times m$ matrix

$$W^{-1} = \operatorname{diag}\left(\frac{1}{w_1}, \frac{1}{w_2}, \ldots, \frac{1}{w_p}\right) \in \mathbb{R}^{n \times m}$$

by transposing W and inverting non-zero elements, so that

$$\begin{aligned} m \le n \quad &\Rightarrow \quad W\, W^{-1} = I_m, \\ m \ge n \quad &\Rightarrow \quad W^{-1}\, W = I_n. \end{aligned}$$

The point here is that if $m < n$ then $W^{-1}\, W$ will have zeroes on the diagonal, and similarly for $W\, W^{-1}$ when $m > n$.

With this notation the solution (A.8) can be rewritten more revealingly as

$$\widetilde{x} = \sum_{j=1}^{\min\{m,n\}} \frac{u_j^T b}{w_j} v_j = A^{-1}\, b \qquad \text{where} \qquad A^{-1} = V\, W^{-1}\, U^T,$$

and the second theorem then says

1. If $m = n$ then A^{-1} is both a left and right inverse to A

$$\begin{aligned} AA^{-1} &= U\, W\, V^T\, V\, W^{-1}\, U^T = U\, W\, I_n\, W^{-1}\, U^T = U\, I_n\, U^T = I_n, \\ A^{-1}A &= V\, W^{-1}\, U^T\, U\, W\, V^T = V\, W^{-1}\, I_n\, W\, V^T = V\, I_n\, V^T = I_n, \end{aligned}$$

   and $\widetilde{x} = A^{-1}\, b$ is the unique solution to $Ax = b$.

2. If $m < n$ then A^{-1} is a right but not left inverse to A

$$\begin{aligned} AA^{-1} &= U\, W\, V^T\, V\, W^{-1}\, U^T = U\, W\, I_n\, W^{-1}\, U^T = U\, I_m\, U^T = I_m, \\ A^{-1}A &= V\, W^{-1}\, U^T\, U\, W\, V^T = V\, W^{-1}\, W\, V^T \ne I_n, \end{aligned}$$

   and $\widetilde{x} = A^{-1}\, b$ is the particular solution to $Ax = b$ with minimum 2-norm.

3. If $m > n$ then A^{-1} is a left but not right inverse to A

$$\begin{aligned} AA^{-1} &= U\, W\, V^T\, V\, W^{-1}\, U^T = U\, W\, W^{-1}\, U^T \ne I_m, \\ A^{-1}A &= V\, W^{-1}\, U^T\, U\, W\, V^T = V\, W^{-1}\, I_m\, W\, V^T = V\, I_n\, V^T = I_n, \end{aligned}$$

   so while $\widetilde{x}$ does not solve $Ax = b$ we can still compute

$$A^{-1}Ax = x = A^{-1}b$$

   which is the least squares best solution to $Ax = b$.

REMARK A.1 When A is an $n \times n$ positive definite matrix the SVD and *eigenvalue decomposition* coincide. To see that, note that A now has a Cholesky decomposition $A = G\,G^T$ where G is a lower triangular matrix. So taking the SVD of G^T

$$\begin{aligned} G^T &= U\,W\,V^T \quad \Rightarrow \quad G = V\,W\,U^T \\ \Rightarrow \quad A &= G\,G^T = V\,W\,U^T\,U\,W\,V^T = V\,W^2\,V^T. \end{aligned}$$

□

A.4.3 Semidefinite programming (SDP)

Karmarkar's 1984 polynomial time algorithm concentrated attention on *interior point methods* for linear programming (LP), and led to extensions to more general convex programs, see [130], [82], [125], [84] and [129]. Instead of tracing the edges of the feasible region to a minimum vertex as in the simplex method, interior point algorithms pursue the *central path* through a convex region to a solution, and so can be generalized to *semidefinite programming* (SDP), in which the variables are symmetric matrices lying in the *convex cone* of *positive semi-definite matrices* (in contrast to real numbers lying in the non-negative orthant in LP). Here are the basic concepts.

Denote the space of *real symmetric* $n \times n$ matrices by

$$\mathbb{S}_n = \left\{X \in \mathbb{R}^{n\times n} : X^T = X\right\},$$

with the *Frobenius inner product* and *F-norm* $\|\cdot\|_F$ defined by

$$X \bullet Y = \text{trace}(X^T Y) = \sum_{i=1}^{n}\sum_{j=1}^{n} X_{ij}Y_{ij} \quad \textit{and} \quad \|X\|_F = \sqrt{X \bullet X}.$$

Note that all matrices in $\mathbb{S}_n$ have real eigenvalues. That is because for any eigenvalue λ with eigenvector X such that $\overline{X}^T X = 1$ clearly A being real implies

$$AX = \lambda X, \quad A\overline{X} = \overline{\lambda X} \quad \Rightarrow \quad \overline{X}^T AX = \lambda, \quad X^T A\overline{X} = \overline{\lambda},$$

and then transposing the second equation (A *symmetric*) $\Rightarrow \overline{X}^T AX = \overline{\lambda}$ and subracting from the first gives $\overline{\lambda} = \lambda$.

A matrix $X \in \mathbb{S}_n$ is *positive semidefinite*, written $X \succeq 0$, if any one of the following equivalent conditions hold:

1. For every vector $w \in \mathbb{R}^n$, $w^T X w \geq 0$.
2. Every eigenvalue of X is nonnegative: $\lambda_i(X) \geq 0$, $i = 1, \ldots, n$.
3. The Cholesky factorization $X = LL^T$, where $L \in \mathbb{R}^n$ is lower triangular, exists.

4. X has a square root $Y \in \mathbb{S}_n$ such that $X = YY^T$.
5. $X \bullet Y \geq 0$ for every $Y \in \mathbb{S}_n$, $Y \succeq 0$.

Importantly, a matrix $X \in \mathbb{S}_n$ is a *covariance* if and only if X is positive semidefinite. Sufficiency comes from Condition-1 because if $\zeta \in \mathbb{R}^n$ is a vector of correlated random variables with $X = \operatorname{cov}\zeta$, then

$$\forall\, w \in \mathbb{R}^n,\ w^T X w = \mathbf{E}\left\{\left(w^T\left[\zeta - \mathbf{E}\zeta\right]\left[\zeta - \mathbf{E}\zeta\right]^T w\right)\right\} = \mathbf{E}\left|w^T\left[\zeta - \mathbf{E}\zeta\right]\right|^2 \geq 0.$$

Necessity comes from Condition-4 because if X is semidefinite matrix, the normal random variable $Y\zeta$ where

$$X = YY^T, \quad \zeta \sim \mathbf{N}\left(0, \mathbf{I}_r\right), \quad r = \operatorname{rank}\left(Y\right)$$

has covariance X.

Semidefinite programming deals with optimization problems of the form

$$\boxed{\begin{array}{rl} \textit{find} & X \in \mathbb{S}_n \\ \textit{to minimize} & C \bullet X \\ \textit{subject to} & A_k \bullet X = b_k \quad k = 1, \ldots, K \\ \textit{and} & X \succeq 0 \end{array}}$$

for given $C \in \mathbb{S}_n$ and $A_k \in \mathbb{S}_n$, $b_k \in \mathbb{R}$ $(k = 1, \ldots, K)$. Note the objective function $C \bullet X$ and the K constraints $A_k \bullet X = b_k$ are all linear functions of the variables X. The dual problem is

$$\boxed{\begin{array}{rl} \textit{find} & y \in \mathbb{R}^m \\ \textit{to maximise} & y^T b \\ \textit{subject to} & C - \sum_{k=1}^{K} y_k A_k \succeq 0 \end{array}}$$

If both primal and dual have strictly feasible points (a stronger condition than is required in linear programming), then the primal and dual objective values are equal at a solution.

LP is just a special case of SDP in which the matrices are diagonal $X = \operatorname{diag}(x)$ and $A_j = \operatorname{diag}(a_j)$ with $x, a_j \in \mathbb{R}^m$ and $X \bullet A_j = a_j^T x$ for $j = 1, .., J$. Note also that several different semidefinite matrices $X_1, X_2, \ldots, X_p$ can be accumulated into one larger block diagonal semidefinite matrix $X = \operatorname{diag}(X_1, X_2, \ldots, X_p)$. A standard problem which explicitly includes both symmetric positive semidefinite matrix variables and vectors of non-negative variables, therefore has form

$$\boxed{\begin{array}{rl} \textit{find} & X \in \mathbb{S}_n \qquad x \in \mathbb{R}^m \\ \textit{to minimize} & C \bullet X + c^T x \\ \textit{subject to} & A_k \bullet X = b_k \quad k = 1, \ldots, K \\ & a_j^T x = \beta_j \quad j = 1, \ldots, J \\ \textit{and} & X \succeq 0 \qquad x \geq 0 \end{array}} \tag{A.9}$$

where X,C and A_k $k = 1,\ldots,m_s$ are symmetric block diagonal matrices, each with p blocks of size $n_1,\ldots,n_p$ and $n = \sum_{i=1}^{p} n_i$, and $a_j \in \mathbb{R}^m$, $\beta_j \in \mathbb{R}$ $j = 1,\ldots,J$.

SDP is closely related to *eigenvalue optimization* problems. If $\lambda_1, \lambda_2, \ldots, \lambda_n$ are the eigenvalues of $X \in \mathbb{S}_n$, the simple objective function

$$\mathbf{I}_n \bullet X = \text{trace}(X) = \sum_{i=1}^{n} \lambda_i,$$

where $\mathbf{I}_n$ is the $n \times n$ identity matrix, can be used to *minimize* the largest eigenvalue of $X \succeq 0$ (and hence the *size* of X) subject to the constraints $A_k \bullet X = b_k$, $k = 1,\ldots,K$, by solving

$$\begin{array}{rl} \textit{find} & X \in \mathbb{S}_n \qquad \zeta \in \mathbb{R} \\ \textit{to minimize} & \zeta \\ \textit{subject to} & \zeta\mathbf{I}_n - X \succeq 0 \\ & A_k \bullet X = b_k \quad k = 1,..,K \\ \textit{and} & X \succeq 0,\ \zeta \geq 0 \end{array}$$

This works because the eigenvalues of $\zeta\mathbf{I}_n - X$ are $\zeta - \lambda_i(X)$, so $\zeta\mathbf{I}_n - X \succeq 0$ is equivalent to $\zeta \geq \lambda_i(X) \geq 0$ for $i = 1,\ldots,n$ (so really the constraint $\zeta \geq 0$ is redundant). This can then be converted into the standard form (A.9) by introducing a *dummy matrix* $D = \zeta I - X \succeq 0$, constructing a $2n \times 2n$ block diagonal matrix diag$\{X,\ D\} \in \mathbb{S}_n$ and then adding n^2 standard equality constraints of type $A_k \bullet X = b_k$ to ensure corresponding elements of D and $\zeta I - X$ are the same.

Now consider the problem of making a variable covariance matrix X *close* (in some sense) to a *target covariance* matrix G. The difference $X - G$ is not generally positive semidefinite, so objective functions like trace$(X - G)$ or largest eigenvalue of $X - G$ are useless. Moreover, neither the Frobenius norm nor its square, can be used directly to force X close to G because $\|X - G\|_F^2 = (X - G) \bullet (X - G)$ is not linear in X. Nevertheless, $X - G$ being symmetric still allows the 2−norm

$$\|X - G\|_2 = \max_{i=1,\ldots,n} |\lambda_i(X - G)|$$

to be minimized by adding two extra constraints and solving

$$\begin{array}{rl} \textit{find} & X \in \mathbb{S}_n \qquad \zeta \in \mathbb{R} \\ \textit{to minimize} & \zeta \\ \textit{subject to} & \zeta\mathbf{I}_n - (X - G) \succeq 0 \\ & \zeta\mathbf{I}_n + (X - G) \succeq 0 \\ & A_k \bullet X = b_k \quad k = 1,..,K \\ \textit{and} & X \succeq 0,\ \zeta \geq 0 \end{array}$$

(convert to the standard form (A.9) by introducing *dummy matrices* $D_1 = \zeta I-(X-G)$ and $D_2 = \zeta I+(X-G)$). This works because the extra constraints imply

$$\begin{aligned}
&\zeta \geq \lambda_i(X-G) \quad \textit{and} \quad \zeta \geq -\lambda_i(X-G) \quad \textit{for} \quad i=1,\ldots,n \\
\Rightarrow \quad &\zeta \geq \max_{i=1,\ldots,n} \{\lambda_i(X-G), -\lambda_i(X-G)\} = \|X-G\|_2.
\end{aligned}$$

Note that the constraint $\zeta \geq 0$ is in fact implied by the positive semidefinite constraints. A more difficult problem is to find the best covariance matrix X with rank(X) specified.

A.5 Some Fourier transform technicalities

Some useful Fourier Transform results are presented in this section, along with some non-rigorous outlines for their derivation.

Using probablistic convention, the *Fourier transform* $\widehat{f}$ of the function f and the *inverse Fourier transform* are defined by the pair

$$\widehat{f}(x) = \int_{-\infty}^{\infty} f(y) \exp(2\pi ixy)\, dy, \qquad f(y) = \int_{-\infty}^{\infty} \widehat{f}(x) \exp(-2\pi ixy)\, dx. \tag{A.10}$$

Standard Fourier theory (involving an isometry plus Cauchy sequences, etc) demonstrates that if $\widehat{f}$ exists then its inverse must be f. Moreover, transforms and inverses are interchangeable in the sense that

$$\int_{-\infty}^{\infty} f(y) \exp(2\pi ixy)\, dy = g(x) \Rightarrow \int_{-\infty}^{\infty} f(x) \exp(-2\pi ixy)\, dx = g(-y).$$

Hence to establish a Fourier transform pair, we need only obtain $\widehat{f}$ from f, or f from $\widehat{f}$.

Two important functions in Fourier analysis are the *Dirac delta function* $\boldsymbol{\delta}(\cdot)$ (defined as a probability measure concentrated at 0, which makes it even), and the *Heaviside function* $\mathbb{I}(\cdot)$ (which *steps* from 0 to 1 at $x=0$)

$$\int_a^b \boldsymbol{\delta}(y)\, dy = \begin{cases} 1 & \text{when} \quad 0 \in (a,b) \\ 0 & \text{when} \quad 0 \notin (a,b) \end{cases}, \quad \mathbb{I}(x) = \begin{cases} 1 & \text{when} \quad x > 0 \\ \frac{1}{2} & \text{when} \quad x = 0 \\ 0 & \text{when} \quad x < 0 \end{cases}. \tag{A.11}$$

Integrating these functions yields expressions for their derivatives

$$\int_{-\infty}^{x} \mathbb{I}(u)\, du = (x)^+ \Rightarrow \frac{d(x)^+}{dx} = \mathbb{I}(x), \quad \int_{-\infty}^{x} \boldsymbol{\delta}(u)\, du = \mathbb{I}(x) \Rightarrow \frac{d\mathbb{I}(x)}{dx} = \boldsymbol{\delta}(x).$$

Note $\mathbb{I}(0) = \frac{1}{2}$ is consistent with $\boldsymbol{\delta}(\cdot)$ being even, and differentiating

$$\begin{aligned}(x)^{+} = x\,\mathbb{I}(x) \quad &\Rightarrow \quad \mathbb{I}(x) = \mathbb{I}(x) + x\boldsymbol{\delta}(x) \\ &\Rightarrow \quad x\boldsymbol{\delta}(x) = 0,\end{aligned}$$

that is, the *zero* in x so to speak *dominates the infinity* in $\boldsymbol{\delta}(x)$ around $x = 0$.

[**1**] Some common Fourier transform pairs are

$\widehat{f}(x)$ $= \int_{-\infty}^{\infty} f(y) \exp(2\pi ixy)\, dy$	$f(y)$ $= \int_{-\infty}^{\infty} \widehat{f}(x) \exp(-2\pi ixy)\, dx$
1	$\boldsymbol{\delta}(y)$
$\mathbb{I}(x)$	$\frac{1}{2}\boldsymbol{\delta}(y) + \frac{1}{2\pi iy}$
$\cos(2\pi ax)$	$\frac{1}{2}[\boldsymbol{\delta}(y+a) + \boldsymbol{\delta}(y-a)]$
$\sin(2\pi ax)$	$\frac{1}{2}i[\boldsymbol{\delta}(y+a) - \boldsymbol{\delta}(y-a)]$
$\delta(x-a)$	$\exp(-2\pi ia\ y)$
$\exp(-2\pi a\ \lvert x\rvert)$	$\frac{1}{\pi}\frac{a}{y^2+a^2}$
$\exp(-a\ x^2)$	$\sqrt{\frac{\pi}{a}}\exp\left(-\frac{\pi^2 y^2}{a}\right)$

[**2**] A useful contour integral is

$$\int_{-\infty}^{\infty} \frac{\exp(2\pi ixy)}{2\pi ix} dx = \mathbb{I}(y) - \frac{1}{2}. \tag{A.12}$$

PROOF Integrating counterclockwise around the contour C consisting of part the real axis $(-R, -r)$, the small semicircle $\{re^{i\theta} : -\pi \le \theta \le 0\}$, part of the real axis (r, R), and the large semicircle $\{Re^{i\theta} : 0 \le \theta \le \pi\}$, we have

$$\oint_C \frac{\exp(2\pi izy)}{2\pi iz} dz = 0.$$

When $y = 0$, breaking down the integral into its component parts, the result follows on letting $r \to 0$ and $R \to +\infty$ in

$$\int_{-R}^{-r} \frac{1}{2\pi ix} dx + \int_{\pi}^{0} \frac{1}{2\pi i\ re^{i\theta}} ire^{i\theta} d\theta + \int_{r}^{R} \frac{1}{2\pi ix} dx + \int_{0}^{\pi} \frac{1}{2\pi i\ Re^{i\theta}} iRe^{i\theta} d\theta = 0.$$

When $y > 0$, on the large semicircle

$$\lim_{R\to\infty} \frac{\exp(2\pi izy)\, dz}{z} = \lim_{R\to\infty} \frac{\exp(2\pi iy\cos\theta\ R - 2\pi y \sin\theta\ R)\, R\ e^{i\theta} d\theta}{R\ e^{i\theta}} = 0,$$

so breaking down the integral into its component parts, the result follows on letting $r \to 0$ in

$$\int_{-\infty}^{-r} \frac{\exp(2\pi ixy)}{2\pi ix} dx + \int_{r}^{\infty} \frac{\exp(2\pi ixy)}{2\pi ix} dx + \int_{\pi}^{0} \frac{\exp\left(2\pi iy\, re^{i\theta}\right)}{2\pi i\, re^{i\theta}} ire^{i\theta} d\theta = 0.$$

When $y < 0$, changing the variable in the integral by setting $s = -x$ and then integrating similarly to the case $y > 0$, clearly gives

$$\int_{-\infty}^{\infty} \frac{\exp(2\pi ixy)}{2\pi ix} dx = -\int_{-\infty}^{\infty} \frac{\exp(-2\pi isy)}{2\pi is} ds = -\frac{1}{2}.$$

□

[**3**] Another useful contour integral is

$$\int_{-\infty}^{\infty} \exp(-2\pi ixy)\, dx = \int_{-\infty}^{\infty} \exp(2\pi ixy)\, dx = \boldsymbol{\delta}(y), \tag{A.13}$$

$$\Rightarrow \qquad \mathbb{I}'(y) = \int_{-\infty}^{\infty} \exp(2\pi ixy)\, dx = \boldsymbol{\delta}(y).$$

PROOF Using the definition $\boldsymbol{\delta}(\cdot)$ and integrating over (a, b) with respect to y, we need to show

$$\int_{-\infty}^{\infty} \left[\frac{\exp(2\pi ixb)}{2\pi ix} - \frac{\exp(2\pi ixa)}{2\pi ix}\right] dx = \begin{cases} 1 & \text{if } \ a < 0 < b, \\ 0 & \text{otherwise} \end{cases},$$

which follows directly from (A.12). □

[**4**] This integral is useful for options:

$$\int_{-\infty}^{s} \exp(2\pi ixy)\, dx = \frac{1}{2}\boldsymbol{\delta}(y) + \frac{\exp(2\pi iy\, s)}{2\pi iy}. \tag{A.14}$$

PROOF Partial differentiation of the integral (A.14) with respect to s gives

$$\exp(2\pi isy) = \exp(2\pi isy),$$

and shows that for $y \neq 0$ the integral must have form

$$\int_{-\infty}^{s} \exp(2\pi ixy)\, dx = e(y) + io(y) + \frac{\exp(2\pi iy\, s)}{2\pi iy},$$

and it remains to identify the functions $e(\cdot)$ and $o(\cdot)$. Equating real and imaginary parts

$$\int_{-\infty}^{s} \cos(2\pi xy)\, dx = e(y) + \frac{\sin(2\pi y\, s)}{2\pi y},$$

$$\int_{-\infty}^{s} \sin(2\pi xy)\, dx = o(y) - \frac{\cos(2\pi y\, s)}{2\pi y},$$

indicates that $e(\cdot)$ must be an even and $o(\cdot)$ an odd function for $y \neq 0$. The changes of variables

$$x = u + \frac{1}{4y}, \quad s = r + \frac{1}{4y}, \quad (y \neq 0),$$

in the second integral, yield

$$\int_{-\infty}^{r} \cos(2\pi y u)\, du = o(y) + \frac{\sin(2\pi y\, r)}{2\pi y},$$

which on comparison with the first integral establishes

$$o(y) = e(y) = 0 \quad \text{for} \quad y \neq 0.$$

Hence the left hand side of (A.14) has form

$$\int_{-\infty}^{s} \exp(2\pi i x y)\, dx = h(y) + \frac{\exp(2\pi i y\, s)}{2\pi i y}, \tag{A.15}$$

where $h(\cdot)$ is an even function concentrated at 0. But reversing the signs of x, y and s in this integral (A.15) gives

$$\int_{s}^{\infty} \exp(2\pi i x y)\, dx = h(-y) - \frac{\exp(2\pi i y\, s)}{2\pi i y},$$

which, on addition to the original integral (A.15) and use of (A.13), yields the result because

$$\int_{-\infty}^{\infty} \exp(2\pi i x y)\, dx = h(y) + h(-y) = 2h(y) = \boldsymbol{\delta}(y).$$

□

[5] An integral along the positive real line yields the reciprocal function:

$$\int_{-\infty}^{\infty} \mathbb{I}(x) \exp(2\pi i x y)\, dx = \frac{1}{2}\boldsymbol{\delta}(y) - \frac{1}{2\pi i y}.$$

PROOF Set $s = 0$ in (A.14) and subtract from (A.13). □

A.6 The chi-squared distribution

For ν a positive integer, the *central* χ^2*-distribution* is defined by the random variable

$$\chi_\nu^2 = \sum_{i=1}^{\nu} X_i^2 \quad \textit{where the} \quad X_i \sim \mathbf{N}(0,1) \quad \textit{are IID},$$

and the *non-central* χ^2*-distribution* by the random variable

$$\chi_\nu^2(\lambda) = \sum_{i=1}^{\nu} (X_i + a_i)^2 \quad \textit{where } \lambda^2 = \sum_{i=1}^{\nu} a_i^2$$

$$\textit{and the} \quad X_i \sim \mathbf{N}(0,1) \quad \textit{are IID.}$$

Note that these definitions can be extended to non-integer values of ν, for example see [46].

1. The distribution of χ_ν^2 is given by

$$\mathbb{P}\left(\chi_\nu^2 \le y\right) = \frac{1}{2^{\frac{\nu}{2}}\Gamma\left(\frac{\nu}{2}\right)} \int_0^y \mathbf{e}^{-\frac{z}{2}} z^{\frac{\nu}{2}-1} dz,$$

   (go to spherical polar coordinates in ν-dimensions to prove that).

2. Only the norm $\lambda = |\mathbf{a}|$ is relevant (and not the components a_i of $\mathbf{a}$) in the distribution of $\chi_\nu^2(\lambda)$ because

$$\mathbb{P}\left(\chi_\nu^2(\lambda) \le y\right) = \mathbf{E}\mathbb{I}\left(\sum_{i=1}^{\nu} (X_i + a_i)^2 \le y\right)$$

$$= \frac{1}{(2\pi)^{\frac{\nu}{2}}} \int_{\mathbb{R}^\nu} \mathbb{I}\left(\sum_{i=1}^{\nu} (x_i + a_i)^2 \le y\right) \exp\left(-\frac{1}{2}\sum_{i=1}^{\nu} x_i^2\right) dv$$

$$= \frac{1}{(2\pi)^{\frac{\nu}{2}}} \int_{\mathbb{R}^\nu} \mathbb{I}\left(\sum_{i=1}^{\nu} x_i^2 \le y\right) \exp\left(-\frac{1}{2}\sum_{i=1}^{\nu} (x_i - a_i)^2\right) dv$$

$$= \frac{1}{(2\pi)^{\frac{\nu}{2}}} \int_{r=0}^{\sqrt{y}} \int_U \exp\left(-\frac{1}{2}\left(r^2 + \lambda^2\right) + \sqrt{\lambda}\mathbf{r}.\mathbf{n}\right) drdS,$$

   where U is the surface of the unit sphere, and $\mathbf{n}$ is a unit vector in the direction of $\mathbf{a} = (a_1, .., a_\nu)$. Hence, rotating axes to a new x_1-axis that lies along $\mathbf{n}$

$$\mathbb{P}\left(\chi_\nu^2(\lambda) \le y\right) = \frac{1}{(2\pi)^{\frac{\nu}{2}}} \int_{r=0}^{\sqrt{y}} \int_U \exp\left(-\frac{1}{2}\left(r^2 + \lambda^2\right) + \sqrt{\lambda}x\right) drdS$$

$$= \mathbf{E}\mathbb{I}\left(\left(X_1 + \sqrt{\lambda}\right)^2 + \sum_{i=2}^{\nu} X_i^2 \le y\right) = \mathbb{P}\left(\left(X_1 + \sqrt{\lambda}\right)^2 + \chi_{\nu-1}^2 \le y\right).$$

3. Cumulative non-central 2-degrees of freedom. The distribution is given by

$$\mathbb{P}\left(\chi_2^2(\lambda) \le y\right) = \mathbb{P}\left(\left(X + \sqrt{\lambda}\right)^2 + Y^2 \le y\right),$$

$$= \frac{1}{2\pi} \int_{r=0}^{\sqrt{y}} \exp\left(-\frac{1}{2}\left(r^2 + \lambda\right)\right) \left\{\int_{\theta=0}^{2\pi} \exp\left(r\cos\theta\sqrt{\lambda}\right) d\theta\right\} rdr,$$

producing the density

$$f(y) = \frac{d}{dy}\mathbb{P}\left(\chi_2^2(\lambda) \le y\right)$$
$$= \frac{1}{4\pi}\exp\left(-\frac{1}{2}(y+\lambda)\right)\int_0^{2\pi}\exp\left(\sqrt{y\lambda}\cos\theta\right)d\theta.$$

Hence during simulation, $\chi_2^2(\lambda)$ may frequently approach 0 because

$$f(0) = \frac{1}{2}\exp\left(-\frac{1}{2}\lambda\right) > 0,$$

indicating $\nu = 2$ is not a good choice for the $\chi_\nu^2(\lambda)$ process.

4. Cumulative non-central 3-degrees of freedom. The distribution is given by

$$\mathbb{P}\left(\chi_3^2(\lambda) \le y\right) = \mathbb{P}\left(\left(X+\sqrt{\lambda}\right)^2 + Y^2 + Z^2 \le y\right),$$
$$= \frac{1}{(2\pi)^{\frac{3}{2}}}\int_{r=0}^{\sqrt{y}}\int_U \exp\left(-\frac{1}{2}\left(r^2+\lambda^2\right) + r\sqrt{\lambda}\cos\theta\right) r^2\sin\theta dr d\theta d\phi,$$

going to polar coordinates and making U the surface of the unit sphere. Hence

$$\mathbb{P}\left(\chi_3^2(\lambda) \le y\right)$$
$$= \frac{2\pi}{(2\pi)^{\frac{3}{2}}}\int_{r=0}^{\sqrt{y}}\int_0^{\pi}\exp\left(\begin{array}{c}-\frac{1}{2}\left(r^2+\lambda^2\right)\\ +r\sqrt{\lambda}\cos\theta\end{array}\right) r^2\sin\theta dr d\theta = P(y,\lambda)$$
$$= \frac{1}{\sqrt{2\pi}\ \lambda}\int_{r=0}^{\sqrt{y}}\exp\left(-\frac{1}{2}\left(r^2+\lambda^2\right)\right)\left[\begin{array}{c}-\exp\left(-r\sqrt{\lambda}\right)\\ +\exp\left(r\sqrt{\lambda}\right)\end{array}\right] r\ dr,$$
$$= \frac{1}{\sqrt{\lambda}}\mathbf{N}'\left(\sqrt{y}+\sqrt{\lambda}\right) - \frac{1}{\sqrt{\lambda}}\mathbf{N}'\left(\sqrt{y}-\sqrt{\lambda}\right)$$
$$+\mathbf{N}\left(\sqrt{y}+\sqrt{\lambda}\right) + \mathbf{N}\left(\sqrt{y}-\sqrt{\lambda}\right) - 1,$$

with density

$$f(y) = \frac{d}{dy}\mathbb{P}\left(\chi_3^2(\lambda) \le y\right)$$
$$= \frac{1}{2\sqrt{\lambda}}\left\{-\mathbf{N}'\left(\sqrt{y}+\sqrt{\lambda}\right) + \mathbf{N}'\left(\sqrt{y}-\sqrt{\lambda}\right)\right\}.$$

So $f(0) = 0$ indicating $\nu = 3$ is a better choice for a $\chi_\nu^2(\lambda)$ to simulate.

A.7 Miscellaneous

A.7.1 Futures contracts

A contract settled *at the margin* (like, for example, a futures contract) is entered into at zero cost, is settled daily as the underlying index G_t changes, and can generally be exited at zero cost. Hence the present value of the payout stream from some time s to any time t generated by G_t

$$\mathbf{E}_0 \left\{ \int_s^t \frac{dG_u}{\beta(u)} \,|\mathcal{F}_s \right\} = 0,$$

must be zero in order for the arrangement to be arbitrage free. So for any s, t

$$M_t = \int_0^t \frac{dG_u}{\beta(u)} \qquad \Rightarrow \qquad \mathbf{E}_0 \{ M_t \,|\mathcal{F}_s \} = M_s$$

making M_t, and any integral with respect to it, a $\mathbb{P}_0$-martingale. In particular the underlying index

$$G_t = \int_0^t \beta(u)\, dM_u$$

must be a $\mathbb{P}_0$-martingale.

Hence the time t value $G_T(t)$ of a futures contract settling to physical $g(T)$ at time T (that is, $G_T(T) = g(T)$) is

$$G_T(t) = \mathbf{E}_0 \{ g(T) \,|\mathcal{F}_t \}.$$

A.7.2 Random variables from an arbitrary distribution

Suppose the random variable Y has density $g(y)$ on the interval $[a, b]$. Then $G : [a, b] \mapsto [0, 1]$ defined by the cumulative density function

$$G(y) = \int_a^y g(u)\, du,$$

is one-one and onto and so invertible. To generate samples of Y, first generate samples X from the uniform distribution $X \sim \mathbf{U}[0, 1]$, and then set $Y = G^{-1}(X)$. The point is that

$$\mathbb{P}[Y \leq y] = \mathbb{P}\left[G^{-1}(X) \leq y\right] = \mathbb{P}[X \leq G(y)] = G(y).$$

A.7.3 Copula methodology

Consider two random variables X and Y whose marginal distributions $F_X(\cdot)$ and $F_Y(\cdot)$ are known. Introduce new random variables $\widetilde{X}$ and $\widetilde{Y}$ defined by

$$\widetilde{X} = \mathbf{N}^{-1} F_X(X) \qquad \widetilde{Y} = \mathbf{N}^{-1} F_Y(Y).$$

By construction both $\widetilde{X}$ and $\widetilde{Y}$ have Gaussian distributions because

$$\begin{aligned}\mathbb{P}\left[\widetilde{X} \leq x\right] &= \mathbb{P}\left[\mathbf{N}^{-1} F_X(X) \leq x\right] = \mathbb{P}\left[X \leq F_X^{-1}(\mathbf{N}(x))\right] \\ &= F_X\left(F_X^{-1}(\mathbf{N}(x))\right) = \mathbf{N}(x).\end{aligned}$$

The *Gaussian copula* hypothesis is that the joint distribution of $\widetilde{X}$ and $\widetilde{Y}$ is also Gaussian with density $Z_2(\bullet, \bullet; \rho)$ in which the correlation ρ can be chosen to satisfy some other modelling requirement (for example, transform the X, Y data and estimate ρ on the derived $\widetilde{X}, \widetilde{Y}$ data). That allows an expectation involving both X and Y to be computed according to

$$\begin{aligned}\mathbf{E} f(X, Y) &= \mathbf{E} f\left(F_X^{-1}\left(\mathbf{N}\left(\widetilde{X}\right)\right), F_Y^{-1}\left(\mathbf{N}\left(\widetilde{Y}\right)\right)\right) \\ &= \mathbf{E} g\left(\widetilde{X}, \widetilde{Y}\right) = \int_{\mathbb{R}^2} g(x, y) Z_2(x, y; \rho)\, dx dy\end{aligned}$$

Obviously, distributions other than Gaussian can be used in a similar way to produce different copulas.

References

[1] Abramowitz, M., Stegun, A. (1972) *Handbook of Mathematical Functions. ISBN 0-486-61272-4.*

[2] Andersen, L. (2000) A simple approach to the pricing of Bermudan swaptions in the multifactor LIBOR market model. *J. Comput. Finance* 3, 5–32.

[3] Andersen, L., Andreasen, J. (2000a) Jump-diffusion processes: volatility smile fitting and numerical methods for option pricing. *Rev. Derivatives Res.* 4, 231–262.

[4] Andersen, L., Andreasen, J. (2000b) Volatility skews and extensions of the Libor market model. *Appl. Math. Finance* 7, 1–32.

[5] Andersen, L., Andreasen, J. (2001) Factor dependence of Bermudan swaptions: fact or fiction? *J. Finan. Econom.* 62, 3–37.

[6] Andersen, L., Brotherton-Ratcliffe, R. (2001) Extended Libor market models with stochastic volatility. *Working paper, Gen Re Financial Products.*

[7] Andersen, L., Piterbarg, V. (2004) Moment explosions in stochastic volatility models. *Working paper.*

[8] Andersen, L., Broadie, M. (2004) Primal-Dual Simulation Algorithm for Pricing Multidimensional American Options. *Management Science Vol 50 No 9 1222-34.*

[9] Antonov, A., Misirpashaev, T. (2006) Markovian projection onto a displaced diffusion generic formulas with application. *Working paper, NumeriX.*

[10] Artzner, P. (1997) On the numeraire portfolio. *Mathematics of Derivative Securities,* M.A.H. Dempster and S.R. Pliska, eds. Cambridge University Press, Cambridge, pp. 53–58.

[11] Artzner, P., Delbaen, F. (1989) Term structure of interest rates: the martingale approach. *Adv. in Appl. Math.* 10, 95–129.

[12] Avellaneda M., Laurence P. (2000) *Quantitative modelling of derivative securities: from theory to practice. ISBN 1-58488-031-7.*

[13] Babbs, S., Webber, N.J. (1997) Term structure modelling under alternative official regimes. *Mathematics of Derivative Securities,* M.A.H. Dempster and S.R. Pliska, eds. Cambridge University Press, Cambridge, pp. 394–422.

[14] Bachelier, L. (1900) Théorie de la spéculation. *Ann. Sci École Norm. Sup.* 17 (1900), 21–86; or *The Random Character of Stock Market Prices,* P.H. Cootner, ed. MIT Press, Cambridge (Mass.) 1964, pp. 17–78.

[15] Bakshi, G.S., Cao C., Chen Z. (1997) Empirical performance of alternative option pricing models. *Journal of Finance 52 pp 2003-2049*

[16] Bakshi, G.S., Madan, D. (1997) A simplified approach to the valuation of options. *Working paper, Dept Finance, University Maryland.*

[17] Balland, P., Hughston, L.P. (2000a) Markov market model consistent with cap smile. *Internat. J. Theor. Appl. Finance* 3, 161–181.

[18] Balland, P., Hughston, L.P. (2000b) Pricing and hedging with a sticky-delta smile. *April Risk Conference, Paris.*

[19] Bates, D. (1996) Jump and stochastic volatility: exchange rate processes implicit in Deutsche mark options. *Review of Financial Studies 9 pp 69-109.*

[20] Baxter, M., Rennie, A. (1996) *Financial Calculus. An Introduction to Derivative Pricing.* Cambridge University Press, Cambridge.

[21] Baz, J., Chacko, G. (2004) *Financial Derivatives. Pricing, Applications and Mathematics.* Cambridge University Press, Cambridge ISBN0-521-81510 X.

[22] Bingham, N.H., Kiesel, R. (2004) *Risk Neutral Valuation: Pricing and Hedging of Financial Derivatives ISBN 1-85233-458-4.*

[23] Black, F., Karasinski, P. (1991) Bond and option pricing when short rates are lognormal. *Finan. Analysts J.* 47(4), 52–59.

[24] Black, F., Scholes M. (1973) The pricing of options and corporate liabilities. *J. Political Econom.* 81, 637–654.

[25] Black, F., Derman, E., Toy, W. (1990) A one-factor model of interest rates and its application to Treasury bond options. *Finan. Analysts J.* 46(1), 33–39.

[26] Bolsa de Mercadorias & Futuros (BMF) Brazilian Exchange English language website <http://www.bmf.com.br/IndexEnglish.asp>

[27] Brace, A., Musiela, M. (1994) A multifactor Gauss Markov implementation of Heath, Jarrow, and Morton. *Math. Finance* 4, 259–283.

[28] Brace, A., Musiela, M. (1997) Swap derivatives in a Gaussian HJM framework. *Mathematics of Derivative Securities,* M.A.H. Dempster and S.R. Pliska, eds. Cambridge University Press, Cambridge, pp. 336–368.

[29] Brace, A., Womersley, R.S. (2000) Exact fit to the swaption volatility matrix using semidefinite programming. *Working paper, University of New South Wales.*

[30] Brace, A., Gątarek, D., Musiela, M. (1997) The market model of interest rate dynamics. *Math. Finance* 7, 127–154.

[31] Brace, A., Dun, T., Barton, G. (2001) Towards a central interest rate model. *Option Pricing, Interest Rates and Risk Management,* E. Jouini, J. Cvitanić and M. Musiela, eds. Cambridge University Press, Cambridge, pp. 278–313.

[32] Brigo, D., Mercurio, F. (2001a) *Interest Rate Models: Theory and Practice.* Springer, Berlin Heidelberg New York. *Theor. Appl. Finance* 5, 427–446.

[33] Brigo, D., Mercurio, F. (2003) Analytical pricing of the smile in a forward LIBOR market model. *Quant. Finance* 3, 15–27.

[34] Broadie, M., Kaya, O. (2004) Exact Simulation of Stochastic Volatility and other Affine Jump Diffusion Processes. *Working paper.*

[35] Broadie, M., Glasserman, P. (1997a) Pricing American-style securities using simulation. *J. Econom. Dynamics Control* 21, 1323–1352.

[36] Carr., P. (2000) A survey of preference free option valuation with stochastic volatility. *April Risk Conference, Paris.*

[37] Da Prato, G., Zabczyk, J. (1992) *Stochastic Equations in Infinite Dimensions.* Cambridge University Press, Cambridge.

[38] Duffie, D., Filipović, D., Schachermayer, W. (2003) Affine processes and applications in finance. *Ann. Appl. Probab.* 13, 984–1053.

[39] Dun, T., Schlögl, E., Barton, G. (2001) Simulated swaption delta-hedging in the lognormal forward LIBOR model. *Inter J Theoretical and Applied Finance 4(4)677-709.*

[40] Dun, T., Schlogl, E. (2005) Cross Currency Basis Modelling. *Working paper, University of Technology, Sydney.*

[41] Dun, T. (2006) Calibration of a Cross Currency Libor Market Model. *Lecture notes.*

[42] Flesaker, B., Hughston, L. (1997) Dynamic models of yield curve evolution. *Mathematics of Derivative Securities,* M.A.H. Dempster and S.R. Pliska, eds. Cambridge University Press, Cambridge, pp. 294–314.

[43] Geman, H., El Karoui, N., Rochet, J.C. (1995) Changes of Numeraire, Changes of Probability Measure and Option Pricing. *J App Prob (32)443-458.*

[44] Glasserman, P., Zhao, X. (1999) Fast Greeks by simulation in forward Libor models. *Journal of Computational Finance 3:5-39.*

[45] Glasserman, P., Zhao, X. (2000) Arbitrage-free discretization of lognormal forward Libor and swaprate models. *Finance & Stochastics 4: 35-68.*

[46] Glasserman, P. (2004) *Monte Carlo Methods in Financial Engineering. ISBN 0-387-00451-3.*

[47] Golub, G.H., Van Loan C.F., (1989) *Matrix Computations. ISBN 0-8018-3772-3.*

[48] Hagan, P.S., Woodward, D.E. (1999a) Equivalent Black volatilities. *Appl. Math. Finance* 6, 147–157.

[49] Hagan, P.S., Woodward, D.E. (1999b) Markov interest rate models. *Appl. Math. Finance* 6, 223-260.

[50] Hagan, P.S., Kumar, D., Lesniewski, A.S., Woodward, D.E. (2002) Managing smile risk. *Wilmott*, September, 84–108.

[51] Haugh, M., Kogan, L. (2004) Pricing American Options: a Duality Approach. *Operations Research 52: 258-270.*

[52] Heath, D.C., Jarrow, R.A., Morton, A. (1992a) Bond pricing and the term structure of interest rates: a new methodology for contingent claim valuation. *Econometrica* 60, 77–105.

[53] Heston, S.L. (1993) A closed-form solution for options with stochastic volatility with applications to bond and currency options. *Rev. Finan. Stud.* 6, 327–343.

[54] Ho, T.S.Y., Lee, S.-B. (1986) Term structure movements and pricing interest rate contingent claims. *J. Finance* 41, 1011–1029.

[55] Hughston, L., (2001) *The New Interest Rate Models.* Risk Books, London.

[56] Hull, J.C. (1994) *Introduction to Futures and Options Markets.* 2nd Prentice-Hall, Englewood Cliffs (New Jersey).

[57] Hull, J.C. (1997) *Options, Futures, and Other Derivatives.* 3rd Prentice-Hall, Englewood Cliffs (New Jersey).

[58] Hull, J., White, A. (1990) Pricing interest-rate-derivative securities. *Review of Financial Studies 3:573-592.*

[59] Hull, J.C., White, A. (1987) The pricing of options on assets with stochastic volatilities. *J. Finance* 42, 281–300.

[60] Hull, J.C., White, A. (1993a) Bond option pricing based on a model for the evolution of bond prices. *Adv. in Futures Options Res.* 6, 1–13.

[61] Hull, J.C., White, A. (1993b) One-factor interest rate models and the valuation of interest rate derivative securities. *J. Finan. Quant. Anal.* 28, 235–254.

[62] Hull, J.C., White, A. (1993c) Efficient procedures for valuing European and American path-dependent options. *J. Derivatives,* Fall, 21–31.

[63] Hull, J.C., White, A. (1994) The pricing of options on interest-rate caps and floors using the Hull-White model. *J. Financial Eng* 2, 287–296.

[64] IMSL, Visual Numerics <www.vni.com>

[65] Jamshidian, F. (1989) An exact bond option pricing formula. *J. Finance* 44, 205–209.

[66] Jamshidian, F. (1997) LIBOR and swap market models and measures. *Finance Stochast.* 1, 293–330.

[67] Jamshidian, F. (2004) Numeraire-invariant option pricing and American, Bermudan and trigger stream rollover. *Version 1.6 Preprint.*

[68] Jarrow, R.A., Yildirim, Y. (2003) Pricing Treasury inflation protected securities and related derivatives using an HJM model. *J. Finan. Quant. Anal.* 38, 337–359.

[69] Joshi, M. (2006) A simple derivation of and improvements to Jamshidian's and Roger's upper bound methods for Bermudan options. *To appear App Math Fin.*

[70] Joshi, M. (2006) Early exercise and Monte Carlo obtaining of tight bounds. *Powerpoint presentation.*

[71] Joshi, M. (2006) Monte Carlo bounds for callable products with non-analytic break costs. *Working paper.*

[72] Karatzas, I., Shreve, S. (1998a) *Brownian Motion and Stochastic Calculus.* 2nd ed. Springer, Berlin Heidelberg New York.

[73] Longstaff, F.,A., Schwartz, E.S. *(1998)* Valuing American options by simulation: A simple least squares approach. *Rev Fin Studies 14:649-676*

[74] Longstaff, F.A., Santa-Clara, E., Schwartz, E.S. (2001a) Throwing away a billion dollars: the cost of suboptimal exercise in the swaptions market. *J. Finan. Econom.* 62, 39-66.

[75] Longstaff, F.A., Santa-Clara, P., Schwartz, E.S. (2001b) The relative valuation of caps and swaptions: theory and empirical evidence. *J. Finance* 56, 2067–2109.

[76] Madan, D.B., Yor, M. (2002) Making Markov martingales meet marginals: with explicit constructions. *Bernoulli* 8, 509–536.

[77] Merton, R.C. (1990) *Continuous-Time Finance.* Basil Blackwell, Oxford.

[78] Miltersen, K., Sandmann, K., Sondermann, D. (1997) Closed form solutions for term structure derivatives with log-normal interest rates. *J. Finance* 52, 409–430.

[79] Musiela, M., Rutkowski, M. (1997) Continuous-time term structure models: forward measure approach. *Finance Stochast.* 1, 261–291.

[80] Musiela, M., Rutkowski, M. (2005) *Martingale methods in financial modelling.* Springer *ISBN 3-540-20966-2.*

[81] NAG, Numerical Algorithms Group <www.nag.co.uk>

[82] Nesterov, Y., Nemirovski, A. (1994) *Interior Point Polynomial Methods in Convex Programming.* SIAM, Philadelphia.

[83] Øksendal, B. (2003) *Stochastic Differential Equations.* 6th edition. Springer, Berlin Heidelberg New York, 2003.

[84] Overton M. L., H. Wolkowic (1997) Semidefinite programming. *Mathematical Programming* 77, 105–110.

[85] Pedersen, M.B. (1998) Calibrating Libor market models. *SimCorp Working Paper.*

[86] Pelsser, A. (2000a) *Efficient Methods for Valuing Interest Rate Derivatives.* Springer, Berlin Heidelberg New York.

[87] Pelsser, A. (2000b) Pricing double barrier options using Laplace transforms. *Finance Stochast.* 4, 95–104.

[88] Pelsser, A., Pietersz, R. (2003) Risk managing Bermudan swaptions in the Libor BGM model. *Working paper.*

[89] Pelsser, A., Pietersz, R. (2004) Swap Vega in BGM: Pitfall and Alternative. *Risk March 91-93.*

[90] Pelsser, A., Pietersz, R., Regenmortel, M. (2004) Fast drift-approximated pricing in the BGM model. *J Comp Fin 8(1) Fall.*

[91] Piterbarg, V.V. (2003a) A stochastic volatility forward Libor model with a term structure of volatility smiles. *Working paper, Bank of America.*

[92] Piterbarg, V.V. (2003b) Mixture of models: A simple recipe for a ... hangover? *Working paper, Bank of America.*

[93] Piterbarg, V.V. (2003c) A practitioner's guide to pricing and hedging callable Libor exotics in forward Libor models. *Working paper.*

[94] Piterbarg, V.V. (2003d) Computing deltas of callable Libor exotics in a forward Libor model. *Working paper.*

[95] Piterbarg, V.V. (2007) Markovian projections for volatility calibration. *April Risk.*

[96] Pliska, S.R. (1997) *Introduction to Mathematical Finance: Discrete Time Models.* Blackwell Publishers, Oxford.

[97] Polypaths <www.polypaths.com>.

[98] WH Press et al (2002) *Numerical Recipes in C++.* ISBN 0521-75033-4.

[99] Protter, P. (2003) *Stochastic Integration and Differential Equations.* 2nd ed. Springer, Berlin Heidelberg New York.

[100] Raise Partner <www.raisepartner.com>

[101] Rebonato, R. (1998) *Interest Rate Option Models: Understanding, Analysing and Using Models for Exotic Interest-Rate Options.* J. Wiley, Chichester.

[102] Rebonato, R. (1999a) On the simultaneous calibration of multifactor lognormal interest rate models to Black volatilities and to the correlation matrix. *J. Comput. Finance* 2, 5–27.

[103] Rebonato, R. (1999b) On the pricing implications of the joint lognormal assumption for the swaption and cap markets. *J. Comput. Finance* 2, 57–76.

[104] Rebonato, R. (2000) *Volatility and Correlation in the Pricing of Equity, FX and Interest-Rate Options.* J. Wiley, Chichester.

[105] Rebonato, R. (2002) *Modern Pricing of Interest-Rate Derivatives: The Libor Market Model and Beyond.* Princeton University Press, Princeton.

[106] Rebonato, R., Joshi, M. (2001) A joint empirical and theoretical investigation of the modes of deformation of swaption matrices: implications for model choice. *Internat. J. Theor. Appl. Finance* 5, 667–694.

[107] Revuz, D., Yor, M. (1999) *Continuous Martingales and Brownian Motion.* 3rd ed. Springer, Berlin Heidelberg New York.

[108] Ritchken, P. (1987) *Options: Theory, Strategy and Applications.* Scott, Foresman and Co., Glenview (Illinois).

[109] Rogers. L.C.G. (2002) Monte Carlo valuation of American options. *Math Finance Vol 12 pp271-286.*

[110] Rozovskii, B.L. (1973) On the Ito-Venttsel formula. *Vestnik Moskovskogo Universiteta, Matematika Vol 28 No 1 pp26-32.*

[111] Rutkowski, M. (2001) Modelling of forward Libor and swap rates. *Option Pricing, Interest Rates and Risk Management,* E. Jouini, J. Cvitanić and M. Musiela, eds. Cambridge University Press, Cambridge, pp. 336–395.

[112] Schlogl, E. (2002) A multicurrency extension of the lognormal interest rate market models. *Finance Stochast.* 6, 173–196.

[113] Schlogl, E (2002) Arbitrage-free interpolation in models of market observable interest rates *Advances in Finance and Stochastics, May*

[114] Schönbucher, P.J. (1999) A market model of stochastic implied volatility. *Phil. Trans. Royal Society* A 357/1758, 2071-2092.

[115] Schoutens, W. (2003) *Lévy Processes in Finance: Pricing Financial Derivatives.* J.Wiley, Chichester.

[116] Scott, L.O. (1997) Pricing stock options in a jump-diffusion model with stochastic volatility and interest rates: applications of Fourier inversion method. *Math. Finance* 7, 413–426.

[117] Seydel, P.J. (2002) *Tools for Computational Finance.* Springer, Berlin Heidelberg New York.

[118] Shiryaev, A.N. (1984) *Probability.* Springer, Berlin Heidelberg New York.

[119] Shiryaev, A.N. (1999) *Essentials of Stochastic Finance: Facts, Models, Theory.* World Scientific, Singapore.

[120] Shreve, S.E. (2004) *Stochastic Calculus for Finance I. The Binomial Asset Pricing Model.* Springer, Berlin Heidelberg New York.

[121] Shreve, S.E. (2005) *Stochastic Calculus for Finance II. Continuous-Time Model.* Springer, Berlin Heidelberg New York.

[122] Singleton, K., Umantsev, L. (2002) Pricing coupon-bond options and swaptions in affine term structure models. *Math. Finance* 12, 427–446.

[123] Sidenius, J. (2000) LIBOR market models in practice. *J. Comput. Finance* 3(3), 5–26.

[124] Taylor, S.J. (1994) Modeling stochastic volatility: a review and comparative study. *Math. Finance* 4, 183–204.

[125] Vandenberghe, L., Boyd, S. (1996) Semidefinite programming, *SIAM Review*, 38, 49–95.

[126] Vasicek, O. (1977) An equilibrium characterisation of the term structure. *J. Finan. Econom.* 5, 177–188.

[127] Venttsel, A.D. (1965) On the equations of the theory of conditional Markov processes *Teoriya veroyatn i ee primenen, X No 2* pp390-393.

[128] Wilmott, P. (1999) *Derivatives: The Theory and Practice of Financial Engineering.* J.Wiley, Chichester New York.

[129] Wolkowicz, H., Saigal, R., Vandenberghe, L. (2000) *Handbook on Semidefinite Programming*, Kluwer.

[130] Wright, S. (1997) *Primal-Dual Interior-Point Methods*, SIAM, Philadelphia.

[131] Wu, L. (2002) Fast at-the-money calibration of LIBOR market model through Lagrange multipliers. *J. Comput. Finance* 6, 33–45.

[132] Wu, L., Zhang, F. (2002) LIBOR market model: from deterministic to stochastic volatility. *Working paper.*

[133] Yor, M. (1992a) *Some Aspects of Brownian Motion. Part I.* Birkhäuser, Basel Boston Berlin.

[134] Yor, M. (2001) *Functionals of Brownian Motion and Related Processes.* Springer, Berlin Heidelberg New York.

Index

For Product Safety Concerns and Information please contact our EU representative GPSR@taylorandfrancis.com
Taylor & Francis Verlag GmbH, Kaufingerstraße 24, 80331 München, Germany

www.ingramcontent.com/pod-product-compliance
Ingram Content Group UK Ltd.
Pitfield, Milton Keynes, MK11 3LW, UK
UKHW021614240425
457818UK00018B/546
* 9 7 8 0 3 6 7 3 8 8 3 7 9 *